NJU SA 2017-2018

南京大学建筑与城市规划学院　建筑学教学年鉴
THE YEAR BOOK OF ARCHITECTURE TEACHING PROGRAM
SCHOOL OF ARCHITECTURE AND URBAN PLANNING NANJING UNIVERSITY
王丹丹 编　EDITOR: WANG DANDAN
东南大学出版社 · 南京　SOUTHEAST UNIVERSITY PRESS, NANJING

建筑设计及其理论
Architectural Design and Theory

张 雷 教 授	Professor ZHANG Lei
冯金龙 教 授	Professor FENG Jinlong
吉国华 教 授	Professor JI Guohua
周 凌 教 授	Professor ZHOU Ling
傅 筱 教 授	Professor FU Xiao
王 铠 副教授	Associate Professo WANG Kai
黄华青 副教授	Associate Professo HUANG Huaqing
钟华颖 讲 师	Lecturer ZHONG Huaying

城市设计及其理论
Urban Design and Theory

丁沃沃 教 授	Professor DING Wowo
鲁安东 教 授	Professor LU Andong
华晓宁 副教授	Associate Professor HUA Xiaoning
胡友培 副教授	Associate Professor HU Youpei
窦平平 副教授	Associate Professor DOU Pingping
唐 莲 副教授	Associate Professor TANG Lian
刘 铨 讲 师	Lecturer LIU Quan
尹 航 讲 师	Lecturer YIN Hang
尤 伟 讲 师	Lecturer YOU Wei

建筑历史与理论及历史建筑保护
Architectural History and Theory, Preservation of Historic Building

赵 辰 教 授	Professor ZHAO Chen
王骏阳 教 授	Professor WANG Junyang
胡 恒 教 授	Professor HU Heng
冷 天 副教授	Associate Professor LENG Tian
王丹丹 讲 师	Lecturer WANG Dandan

建筑技术科学
Building Technology Science

鲍家声 教 授	Professor BAO Jiasheng
吴 蔚 副教授	Associate Professor WU Wei
郜 志 副教授	Associate Professor GAO Zhi
童滋雨 副教授	Associate Professor TONG Ziyu
梁卫辉 副教授	Associate Professor LIANG Weihui
施珊珊 讲 师	Lecturer SHI Shanshan

南京大学建筑与城市规划学院建筑系
Department of Architecture
School of Architecture and Urban Planning
Nanjing University
arch@nju.edu.cn   http://arch.nju.edu.cn

教学纲要
EDUCATIONAL PROGRAM

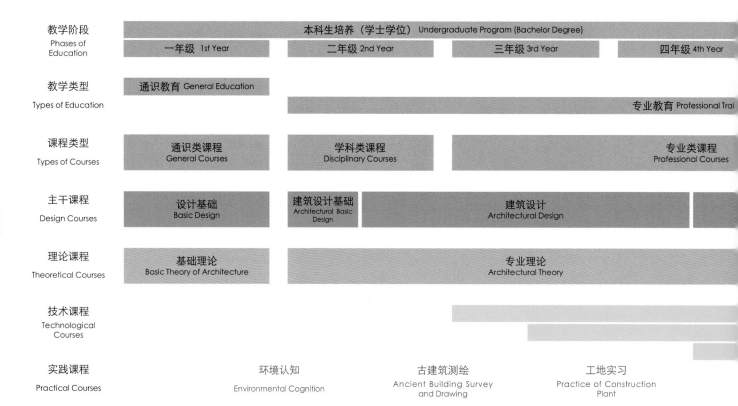

研究生培养（硕士学位）Graduate Program (Master Degree)			研究生培养（博士学位）Ph. D. Program
一年级 1st Year	二年级 2nd Year	三年级 3rd Year	

学术研究训练 Academic Research Training	

学术研究 Academic Research

建筑设计研究 Research of Architectural Design	毕业设计 Thesis Project	学位论文 Dissertation	学位论文 Dissertation
专业核心理论 Core Theory of Architecture	专业扩展理论 Architectural Theory Extended	专业提升理论 Architectural Theory Upgraded	跨学科理论 Interdisciplinary Theory

建筑构造实验室 Tectonic Lab
建筑物理实验室 Building Physics Lab
数字建筑实验室 CAAD Lab

生产实习　　　　生产实习
Practice of Profession　　Practice of Profession

课程安排
CURRICULUM OUTLINE

	本科一年级 Undergraduate Program 1st Year	本科二年级 Undergraduate Program 2nd Year	本科三年级 Undergraduate Program 3rd Year
设计课程 Design Courses	设计基础 Basic Design	建筑设计基础 Architecture Basic Design 建筑设计（一） Architectural Design 1 建筑设计（二） Architectural Design 2	建筑设计（三） Architectural Design 3 建筑设计（四） Architectural Design 4 建筑设计（五） Architectural Design 5 建筑设计（六） Architectural Design 6
专业理论 Architectural Theory		建筑导论 Introductory Guide to Architecture	建筑设计基础原理 Basic Theory of Architectural Design 居住建筑设计与居住区规划原理 Theory of Housing Design and Residential Planning 城市规划原理 Theory of Urban Planning
建筑技术 Architectural Technology	理论、材料与结构力学 Theoretical, Material & Structural Statics Visual BASIC程序设计 Visual BASIC Programming	CAAD理论与实践（一） Theory and Practice of CAAD1	建筑技术（一） 结构与构造 Architectural Technology 1: Structure & Construction 建筑技术（二） 建筑物理 Architectural Technology 2: Building Physics 建筑技术（三） 建筑设备 Architectural Technology 3: Building Equipment
历史理论 History Theory		外国建筑史（古代） History of World Architecture (Ancient) 中国建筑史（古代） History of Chinese Architecture (Ancient)	外国建筑史（当代） History of Western Architecture (Modern) 中国建筑史（近现代） History of Chinese Architecture (Modern)
实践课程 Practical Courses		古建筑测绘 Ancient Building Survey and Drawing	工地实习 Practice of Construction Plant
通识类课程 General Courses	数学 Mathematics 语文 Chinese 思想政治 Ideology and Politics 科学与艺术 Science and Art	社会学概论 Introduction of Sociology	
选修课程 Elective Courses		城市道路与交通规划 Planning of Urban Road and Traffic 环境科学概论 Introduction of Environmental Science 人文科学研究方法 Research Method of the Social Science 管理信息系统 Management Operating System 城市社会学 Urban Sociology	人文地理学 Human Geography 中国城市发展建设史 History of Chinese Urban Development 欧洲近现代文明史 Modern History of European Civilization 中国哲学史 History of Chinese Philosophy 宏观经济学 Macro Economics

本科四年级	研究生一年级	研究生二、三年级
Undergraduate Program 4th Year	Graduate Program 1st Year	Graduate Program 2nd & 3rd Year
建筑设计（七） Architectural Design 7 建筑设计（八） Architectural Design 8 本科毕业设计 Graduation Project	建筑设计研究（一） Design Studio 1 建筑设计研究（二） Design Studio 2 数字建筑设计 Digital Architecture Design 联合教学设计工作坊 International Design Workshop	专业硕士毕业设计 Thesis Project
城市设计理论 Theory Urban Design	城市形态研究 Study on Urban Morphology 现代建筑设计基础理论 Preliminaries in Modern Architectural Design 现代建筑设计方法论 Methodology of Modern Architectural Design 景观都市主义理论与方法 Theory and Methodology of Landscape Urbanism	
建筑师业务基础知识 Introduction of Architects' Profession 建设工程项目管理 Management of Construction Project CAAD理论与实践（二） Theory and Practice of CAAD2	材料与建造 Materials and Construction 中国建构（木构）文化研究 Studies in Chinese Wooden Tectonic Culture 计算机辅助技术 Technology of CAAD GIS基础与运用 Concepts and Application of GIS	
	建筑理论研究 Study of Architectural Theory	
生产实习（一） Practice of Profession 1	生产实习（二） Practice of Profession 2	建筑设计与实践 Architectural Design and Practice
景观规划设计及其理论 Theory of Landscape Planning and Design 地理信息系统概论 Introduction of GIS 欧洲哲学史 History of European Philosophy 微观经济学 Micro Economics 建筑技术中的人文主义 The Technology of Humanism in Architecture 建筑节能与绿色建筑 Building Energy Efficiency and Green Building Design	建筑史研究 Studies in Architectural History 建筑节能与可持续发展 Energy Conservation & Sustainable Architecture 建筑体系整合 Advanced Building System Integration 规划理论与实践 Theory and Practice of Urban Planning 景观规划进展 Development of Landscape Planning	

1-129 年度改进课程 WHAT'S NEW

2
设计基础（二）
BASIC DESIGN 2

52
本科毕业设计：互动建筑原型数字化设计与建造
GRADUATION PROJECT: INTERACTIVE ARCHITECTURAL PROTOTYPE DIGITAL DESIGN AND CONSTRUCTION

16
建筑设计（二）：独立居住空间设计
ARCHITECTURAL DESIGN 2: INDEPENDENT LIVING SPACE DESIGN

60
本科毕业设计：意大利旧建筑改造设计
GRADUATION PROJECT: RESTORATION OF HISTORIC ARCHITECTURE IN ITALY

28
建筑设计（三）：乡村小型家庭旅馆扩建
ARCHITECTURAL DESIGN 3: EXTENSION OF A COUNTRY HOUSE AS HOLIDAY INN

38
建筑设计（五+六）：城市建筑：大型公共建筑设计
ARCHITECTURAL DESIGN 5 & 6: URBAN ARCHITECTURE: COMPLEX BUILDING

72
研究生国际教学交流计划课程
THE INTERNATIONAL POST-GRADUATE TEACHING PROGRAM

112
建筑设计研究（二）：多层木结构建筑设计
DESIGN STUDIO 2: DESIGN OF A MULTI-STOREY TIMBER STRUCTURE BUILDING

122
建筑设计研究（三）：异质类型：建筑、基础设施和地景（1）相地
DESIGN STUDIO 3: HETEROTYPE: ARCHITECTURE, INFRASTRUCTURE, LANDSCAPE (1) SITE DESCRIPTION

147—197 附录 APPENDIX

147—159 建筑设计课程 ARCHITECTURAL DESIGN COURSES

161—163 建筑理论课程 ARCHITECTURAL THEORY COURSES

165—167 城市理论课程 URBAN THEORY COURSES

169—171 历史理论课程 HISTORY THEORY COURSES

173—175 建筑技术课程 ARCHITECTURAL TECHNOLOGY COURSES

177—189 回声——来自毕业生的实践 ECHO—FROM PRACTICES OF GRADUATES

191—197 其他 MISCELLANEA

130
建筑理论课程："建筑史方法"与李渔的《十二楼》
ARCHITECTURAL THEORY COURSES: "ARCHITECTURAL HISTORY METHODOLOGY" AND LI YU'S *TWELVE PAVILIONS*

年度改进课程
WHAT'S NEW

设计基础（二）
BASIC DESIGN 2

鲁安东 唐莲 尹航 孟宪川 黄华青

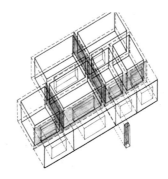

教学内容
（1）采用课内多模块选修制度。整个课程分为3个阶段，每个阶段时长5周。
（2）学生分成A班（小班）和B班（大班）。A班每班20人，共2个班，采用设计教学（studio）的动手实操形式。B班1个班，60~80人，采用大课授课（lecture）形式。
（3）A班共有6个教学模块（每阶段2个），A班学生在学期期间共选修其中3个模块。6个模块为不同的设计练习，具体内容由教师进行设计。

教学要点
（1）A类模块
本学期的6个A类教学模块突出美学素质培养和设计思维训练，重在培养设计感和设计兴趣，可包含适量动手环节，但不宜强调技术训练和工作量。各模块应富有趣味，特色鲜明，成果明确。
A类教学模块主要内容为 感知：培养感受与思维的协同，包括对空间的感受、对身体的感受、对氛围的感受等；表达：培养表达与思维的协同，掌握不同媒介、可视化手段，让学生具备运用表达手段进行想象和思考的能力。
（2）B班
B班由16次大课构成，由4位老师分别完成。内容为建筑学介绍、建筑鉴赏、建筑历史等带有通识性质的讲课。16次大课中：第1周为课程介绍，2~15周为讲课，另有1周为A班和B班集体汇总成果展示（评A班成果）。

教师工作量
每位教师可选教授2个A模块+2次大课，或者1个A模块+5次大课。

Teaching Content
(1) A multi-module elective course system in the class is implemented. The entire course is divided into 3 phases, each of which lasts 5 weeks.
(2) The students are assigned into Class A and Class B. Class A has 2 classes, each of them containing 20 pupils. A studio classroom featured with hands-on activities and active learning is adopted in Class A. Class B, which has a capacity of 60~80 pupils, adopts lectures as a teaching method.
(3) Class A has 6 teaching modules (2 for each phase. Among which 3 of them are required by pupils of Class A as elective courses during the semester. The 6 modules are studio practices distinct from each other, and specific contents of these modules are to be designed by the instructor.

Teaching Essential
(1) Modules used in Class A
The 6 Class A teaching modules of this semester highlight the trainings on aesthetics and design thinking, focusing on cultivating inclination to design as well as interest in design, which may include appropriate operational procedures, without putting too much emphasis on technical training and workload. Each module should be designed interesting, distinctive and with a clear goal.
Main contents of Class A teaching modules include: Sensing: Cultivate the synergy between feeling and thinking, including the feeling of space, the feeling of the body and the feeling of the atmosphere; Expressing: Cultivate the synergy between expression and thinking, and the ability to manage different media and visualization methods, empowering the students the ability to imaging and thinking by means.
(2) Class B
Class B consists of 16 lectures, to be assigned to 4 instructors. The content of the lectures is on general knowledge such as introduction to architecture, appreciation of architecture as well as history of architecture. Among the 16 lectures: the course in Week 1 is the introduction to the course, the main courses are in Week 2~15, and an achievement showcase course for Class A and B is in the last week (achievements of Class A will be evaluated).

Instructor Workload
The instructors can each choose to teach 2 A modules and 2 lectures, or 1 A module and 5 lectures.

| 感知（转化能力训练） | 分析（制图能力训练） | 创造（动手能力训练） |
| 5周（个人作业） | 5周（个人作业） | 5周（小组作业） |

A1 鲁安东 电影空间

B1 尹航 城市空间认知

C1 鲁安东 尹航 园林剧场

A2 唐莲 度量空间

B2 黄华青 日常空间（聚落）认知

C2 孟宪川 结构造型

A3 黄华青 书写空间

B3 孟宪川 材料空间认知

C3 唐莲 折纸包裹

介绍课（建筑：鲁安东）　　古镇游　（规划：申明锐）　　园林游（建筑：鲁安东）　　作业展

课程结构说明　The curriculum structure

设计基础（二）BASIC DESIGN 2

A3：书写空间
A3: WRITING SPACE
黄华青

教学目标
培养基于历史文本的美学鉴赏和空间想象能力。

教学内容
"书写空间"模块从明清文人生活典籍出发，引导学生在园林、建筑、艺术、器物之间感受传统生活美学，培养基于历史文本的美学鉴赏和空间想象能力。教学过程历时5周（含评图1周），包括3个练习：
（1）独立完成一组园林空间观察和文本转换；
（2）两人一组，提取园林场景的一个美学元素，结合典籍进行鉴赏分析；
（3）四人一组，基于描述文人生活场景的文本，完成一件场景空间的拼贴作品。

教学进度
第一周：讲课：《长物志与文人生活空间》，针对《长物志与文人生活空间》中的空间描述文本，进行翻译、解读、转译。
第二周：讲课：《园林空间美学的元素》。
第三周：讲课：《场景模型的制作》。
第四周：（1）汇报模型制作方案；（2）指导模型制作。

Teaching Objective
Cultivate historical text-based aesthetic appreciation and space imagination ability.

Teaching Content
"Writing space" module starts from the living records of literati in Ming and Qing dynasties, guides students to feel traditional aesthetics of life in garden, architecture, art, utensil, and cultivates their historical text-based aesthetic appreciation and space imagination ability. The teaching lasts for 5 weeks (including 1 week of drawing evaluation), including 3 exercises:
(1) Independently complete a group of garden space observation and text conversion;
(2) Two students in a group, extract an aesthetic element from garden scene, combine with historical records for appreciation and analysis;
(3) Four students in a group, based on text that describes living scene of literati, complete a collage work of scene space.

Schedule of the Design
First week: Lecture *Treatised on Superfluous Things and Living Space of Literati*, translate, interpret space description text in *Treatise on Superfluous Things and Living Space of Literati*, and translate into a text.
Second week: Lecture: *Aesthetic Element of Garden Space*.
Third week: Lecture: *Making of Scene Model*.
Fourth week: (1) Report model making plan; (2) Guide model making.

鉴赏

阁

作房闼者,须回环窈窕;供登眺者,须轩敞宏丽,藏书画者,须爽垲高深。此其大略也……阁作方样者,四面一式。楼前忌有露台卷篷,楼板忌用砖铺。盖既名楼阁,必有定式。若复铺砖,与平屋何异?

玉兰

宜种厅事前,对列数株。花时,如玉圃琼林,最称绝胜。别有一种紫者,名木笔,不堪与玉兰作婢,古人称辛夷,即此花。

堂

堂之制,宜宏敞精丽,前后须层轩广庭,廊庑俱可容一席。四壁用细砖砌者佳,不则竟用粉壁。梁用球门,高广相称。层阶俱以文石为之,小堂可不设窗槛。

桂

丛桂开时,真称香窟,宜辟地二亩,取各种并植,结亭其中,不得颜以「小山」等语,更勿以他树杂之。

松

取柽子松植堂前广庭,或广台之上,不妨对偶。斋中宜植一株,下用文石为台,或太湖石为栏俱可。

小池

阶前石畔凿一小池,必须湖石四围,泉清澈可见底。中畜朱鱼、翠藻,游泳可玩。四周树野藤、细竹,能掘地稍深,引泉脉,四周树野藤、细竹,能掘地稍深,引泉脉

拟像

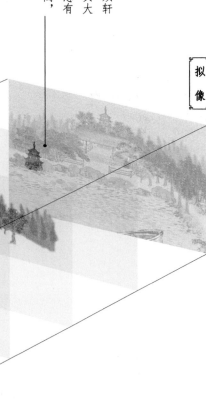

书写瞻园

自观鱼亭后回廊拾级西上,廊边梅花怒放,及至北侧廊级西上,松柏生长,迎春与玉兰争艳,水池对面的春波亭清晰可见,有数游人于亭中观赏池鱼,抱石轩纯朴典雅。登入楹台,中有几案供品茗休息,穿都坊北侧而门,门亮门同,水池亭台若隐若现。目力所及处棒生春色,映在假山和春柳之间,此亭皆入眼底,南有洞门、门外假山正立,园林之色,更一步,又于恍惚间别有洞天,沿回廊南向,花木丛生,偶有几只猫身姿婀娜,山横卧,徐行至板桥,桥倒假于石桥与假山间迂道打闹,好不快活。穿板桥而过即可达春波亭。

入瞻园后,两侧增堂上呈"吉祥""如意"字样。由左侧门入状园,绿得见园林。远山出翠色,山畔临廊,叫人生得一种亲近自然之意,前行数步,水面放清聚,此时山石湖多,映得水面是灰白之色,略带青绿之缘。抬阶穿过水面,大湖石纹行走多不便,恍德穿行于山石间,上下不足,时在晌午,至左右环石凸凹,大可感受所谓科明春寒,至于高处,又是春光明媚,一块近处,前有蒙水,四周绿植环绕,亦有入此观赏水中锦鲤,再行看长廊之地势处,围墙下的石中水汨汨而出,叫人欣喜。一眼则是居曾,其中盆栽可供赏玩,瞻园之色也在于山水草木,唯那些草木有些爱风,繁杂了些笑。

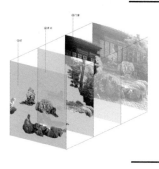

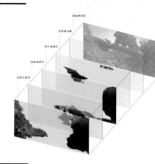

走出曲静的长廊,来到石子小路上,入眼的便是这座颇有气势的假山。与整座瞻园的儒雅幽静不同的是,这座假山自上而下飞溅的水流,显现出一种婆的气势,与旁边曲静的小路相互衬托,也为水池由折奇石环抱,空间的设计给瞻山带来了一种虚怀着谷的气势,加上中间石桥的点缀,实为豪笔,环绕假山的草木,不经意的一笔,添上了豪多的生机。假山不凡的气度中带来了一丝和之感,也为假山带来了更多的欣赏空间,使得其不显得草调无味。

坐静妙宣秀,忽闻泉声深深,遥数鸳啼哗啼,潺潺寻声,目随聆传,然开朗。悬泉飞湍,魏岩漏生奇,碧潭如镜,深窗曲折探幽,杂花楚苍,峰接环翠,暗香疏影,桃李成荫,千浮绿蓝,乌栖懒但闻声,鱼戏上浮桥,见蜻蜒,深润,不闻两股战战,俯堂依长廊,摇清凤真翼,但建美景良辰。千般晨光,万种水色,叹赏造之精巧,恍如纳天然风月入怀,恰似佛古时君子共游。

设计基础（二）BASIC DESIGN 2

B3: 材料空间认知
B3: MATERIAL SPACE COGNITION

孟宪川

教学目标

培养对材料的感性与理性的理解能力，建立对材料空间进行日常使用表述和科学原理阐述的能力。

教学内容

"材料空间认知"的教学历时5周（含评图1周），教学内容包括模型制作、图解绘制、文字表述等。通过对日常材料空间的认知，训练观察材料、科学理解材料和组织材料空间的方式，培养材料空间的感知能力。教学过程包括3个阶段的练习：悬挂材料空间认知（1周），拱材料空间认知（1周），悬挂/拱材料空间设计（2周）。

教学进度

第六周：（1）介绍；（2）讲课：日常材料形成的空间（孟宪川）；（3）课上练习（认知材料空间，利用准备的悬挂材料做模型）；（4）讲课：悬挂材料的图解方法。
第七周：（1）课后作业讲评：电子版；（2）课上练习（结合课堂空间，利用准备的拱材料做模型）；（3）讲课：拱材料的图解方法（孟宪川）。
第八周：（1）课后作业讲评：电子版；（2）讲课：悬挂和拱空间的结构与形态设计（孟宪川）。
第九周：（1）课后作业讲评：电子版。

Teaching Objective

Cultivate sensible and rational material understanding ability, establish ability of daily use expression and scientific principle expatiation of material space.

Teaching Content

Teaching of "material space cognition" lasts for 5 weeks (including 1 week of drawing evaluation), teaching content includes model making, graphic drawing, text description, etc. through cognition of daily material space, train material observation, scientific understanding of material and modes of organizing material space, cultivate material space sensing ability. The teaching includes exercise of 3 stages: suspension material space cognition (1 week), arch material space cognition (1 week), suspension/arch material space design (2 weeks).

Schedule of the Design

Sixth week: (1) Warm Up & Introduction; (2) Lecture: Space Formed by Daily Material (Meng Xianchuan); (3) Class exercise (cognize material space, use prepared suspension material to make models); (4) Lecture: Graphic Method of Suspension Material.
Seventh week: (1) Schoolwork comment: electronic version; (2) Class exercise (combine with class space, use prepared arch material to make models); (3) Lecture: Graphic Method of Arch Material (Meng Xianchuan).
Eighth week: (1) Schoolwork comment: electronic version; (2) Lecture: Structure and Form Design of Suspension and Arch Space (Meng Xianchuan).
Ninth week: (1) Schoolwork comment: electronic version.

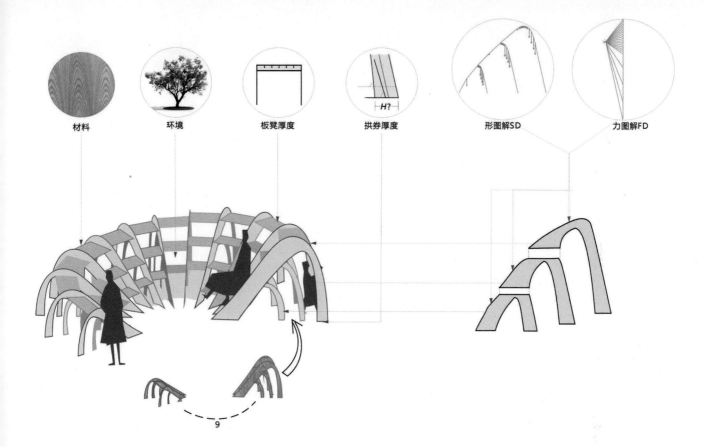

方案构思
Scheme conception

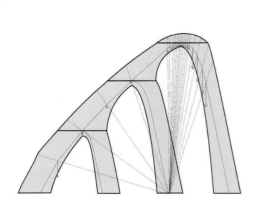

方案生成
Scheme generation

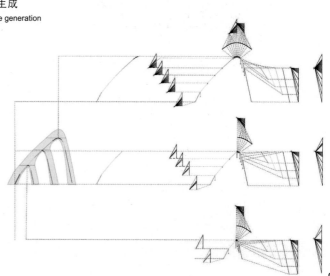

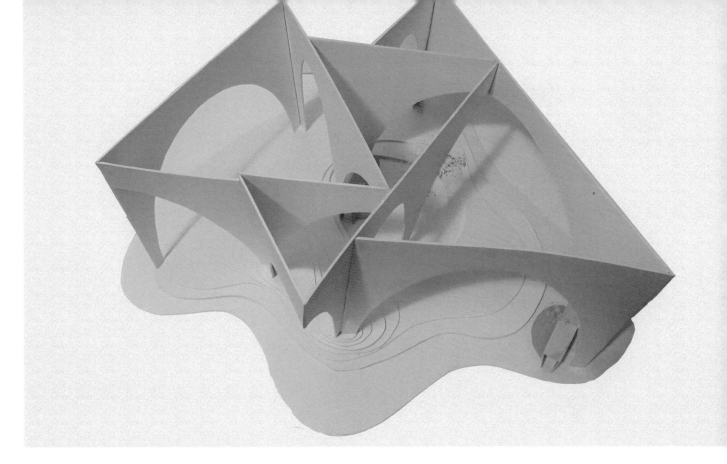

方案生成
Scheme generation

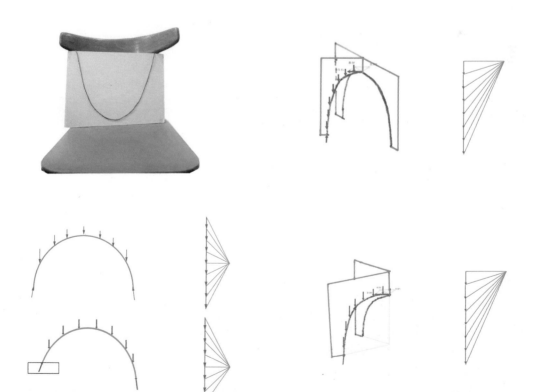

设计基础（二） BASIC DESIGN 2
C3：折纸包裹
C3: PAPER FOLDING WRAP
唐莲

教学目标
理解形式塑造机制，理解形式与材料、构件、工艺的关系。

教学内容
"折纸包裹"的教学历时5周，要求用折纸对身体部位进行包裹，完成设计与制作。课程可以理解为基于身体（场地）的形式操作，教学的主要内容是形式设计的逻辑与方法，其中折纸作为实现形式的技术与媒介。为此，在整个教学过程中设置了3个阶段的练习，并开展相应的讲座来指导与配合练习。这3个阶段分别为：折纸单元基础练习（1周）、折纸单元变形与组合研究（1周）以及折纸包裹空间的设计（3周）。

教学进度
第十一周：介绍。
第十二周：（1）课后作业讲评，纸球，扫描合成图（电子版）；（2）讲课：折纸单元的变形与组合；（3）课上练习：折纸单元变形及组合练习。
第十三周：（1）课后作业讲评：扫描合成图（电子版）；（2）讲课：身体尺度与空间包裹；（3）课上练习：选择身体部位，进行几何分析。
第十四周：（1）课后作业讲评：包裹部位的几何构成图，设计概念图（电子版）及半成品；（2）课上练习：根据意见修改设计，继续完善设计。
第十五周：（1）课后作业讲评；（2）讲课：图示表达Ⅱ。

Teaching Objective
Understand form building mechanism, understand relationship between form and material, component, craft.

Teaching Content
The teaching of "paper folding wrap" lasts for 5 weeks. It is required to use paper folding to wrap body part, complete design and making. The course can be understood as body (site)-based form operation. The main teaching content is logic and method of form design, in which paper folding is used as the technique and medium to realize form. Therefore, exercise of 3 stages is set in the whole teaching, and relevant lecture is provided to guide and cooperate with the exercise. The 3 stages are respectively paper folding unit basic exercise (1 week), paper folding unit deformation and combination research (1 week), and paper folding wrap space design (3 weeks).

Schedule of the Design
Eleventh week: Introduction.
Twelfth week: (1) Schoolwork comment: paper ball, scan composite graph; (2) Lecture: Deformation and Combination of Paper Folding Unit; (3) Class exercise: paper folding unit deformation and combination exercise.
Thirteenth week: (1) Schoolwork comment: scan composite graph (electronic version); (2) Lecture: Body Dimension and Space Wrap; (3) Class exercise: select body part for geometric analysis.
Fourteenth week: (1) Schoolwork comment: geometric composition of wrapped part, design concept drawing (electronic version) and semi-finished work; (2) Class exercise: change design according to the opinion, continue to improve design.
Fifteenth week: (1) Schoolwork comment; (2) Lecture: Graphic Expression II.

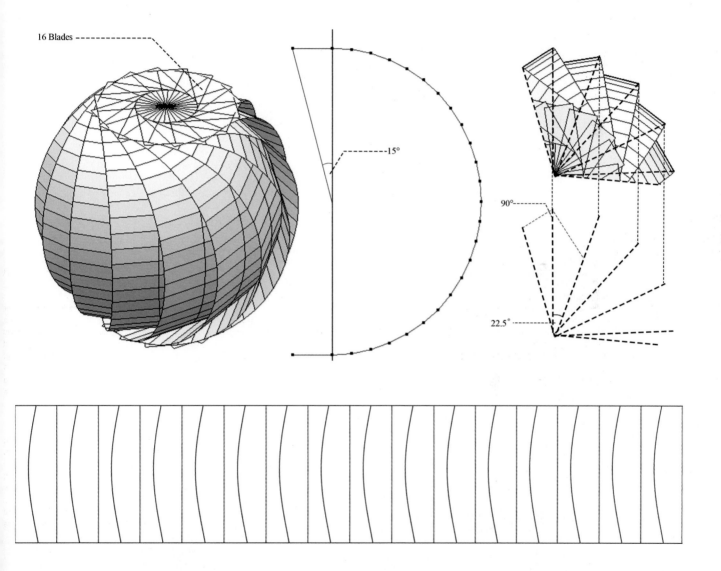

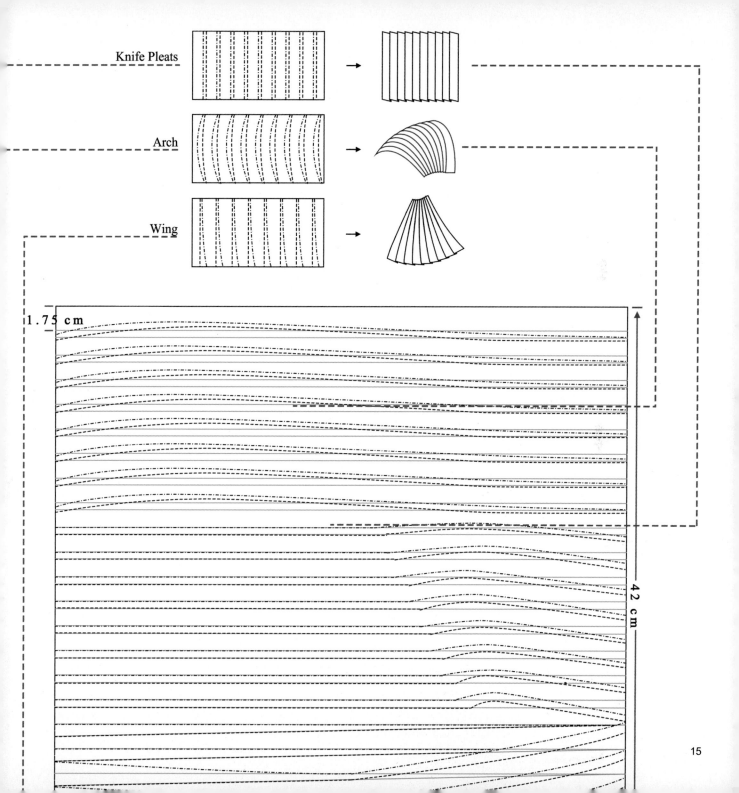

建筑设计（二）ARCHITECTURAL DESIGN 2

独立居住空间设计
INDEPENDENT LIVING SPACE DESIGN

冷天 刘铨 王丹丹

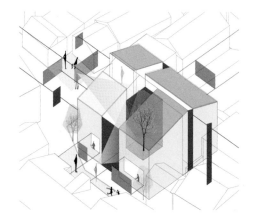

教学目标

综合运用建筑设计基础课程中的知识点，完成一个小型独立居住空间的设计方案。训练的重点一是内部空间的整合性设计，二是通过真实有效的建构设计将空间效果清晰地予以呈现。

教学要点

（1）场地与界面：本次设计场地是南京老城内的真实建筑地块，面积在200 m²左右，单面或相邻两面临街，周边为1~2层的传统民居。要求从场地原有界面出发来考虑新建建筑的形体、布局及其最终的空间视觉感受。

（2）功能与空间：使用者的不同功能需求是建筑空间生成的主要动因，也是建筑设计要解决的基本问题。本练习的功能设定为小型家庭独立居住空间，附设一个小型文化沙龙。家庭主要成员包括一对年轻夫妇（均具有设计行业背景）和1位未成年子女（7~12岁）。新建建筑面积约160~200 m²（上下浮动5%以内），建筑高度≤8 m（指内部可用空间总高，不包括女儿墙，不设地下空间），其中居住空间应包括起居室、餐厅、主卧室、儿童房、多功能房（含工作空间和客床）、厨房、餐厅、卫生间（1~2处）以及必要的储藏空间，并应同时满足文化沙龙的功能，其中包括阅读区、文化展示和沙龙空间、储藏区等。文化沙龙应考虑与住宅空间合理结合。

（3）流线与出入口：建筑内部各功能空间不仅需要合理的水平、垂直交通来相互沟通与联系，还要考虑与场地周边环境条件的合理衔接，如街道界面的连续性、出入口位置的选择与退让处理、周边建筑外墙界面（包括其上的外窗）对新建建筑的影响、建筑之间的间距和视线干扰、日照的合理使用等。新建筑的内部楼梯必须符合现行规范（踏面宽度25~30 cm，踏步高15~20 cm，连续踏步数量不得超过18级等）。

（4）尺度与感知：建筑内部的空间是供人来使用的，因此建筑中的各功能空间的尺度，都必须以人体作为基本的参照和考量，并结合人体的各种行为活动方式，来确定合理的建筑空间尺寸。在空间形式处理中注意通过图示表达理解空间构成要素与人的空间体验之间的关系，主要包括尺度感和围合感。

（5）结构与构造：根据具体的空间特点选择合理有效的支撑体系和包裹体系，通过一系列较为全面的材料、结构和构造设计，完成空间的限定和视觉效果的展示。

教学安排

本次设计课程共7周（2018.05.04—2018.06.22）

第一周：调研场地，分组制作1:100场地模型（底座60 cm×60 cm×5 cm），用工作模型构思初步方案，通过照片拼贴、沿街透视研究场地和界面的关系与效果。

第二周：用1:50手绘平、立、剖面图纸，综合研究功能与空间、流线与尺度。

第三周：确定基本设计方案，推进剖、立面设计，利用工作模型辅助（集中挂图点评）。

第四周：深化设计方案并细化推敲各设计细节，并建模研究内部空间效果。

第五周：深化1:50图纸，开始照片拼贴效果图的制作（集中挂图点评）。

第六周：制作1:50剖透视和各分析图，制作1:50大比例模型（集中挂图点评）。

第七周：整理图纸，排版，并完成课程答辩。

成果要求

A1图纸3张，纸质表现模型1个（比例1:50），工作和研究模型若干。图纸内容应包括：

（1）总平面图（1:200），各层平面图、纵横剖面图和主要立面图（1:50），内部空间组织及建构表达的剖透视图1张（1:50）。

（2）设计说明和主要技术经济指标（用地面积、建筑面积、容积率、建筑密度）。

（3）表达设计意图和设计过程的分析图（体块生成、功能分析、流线分析、结构体系等）。

（4）纸质模型照片与电脑效果图、照片拼贴等。

Teaching Objective

Comprehensively apply the knowledge in architectural design basic course, complete a small independent living space design project. The training focus is integration design of internal space, and clearly showing space effect through real and effective tectonic design.

Teaching Essential

(1) Site and interface: The design site is a real architectural plot in Nanjing old city, with the area of 200 m², facing the street by one side or adjacent two sides, with traditional residence of 1~2 storeys surrounding. Students are mainly required to consider designing architectural form, layout and final space visual feeling from the

perspective of original site interface.

(2) Function and space: Different function requirements of user are the main motivation for generating building space, and also the basic issue to solve in architectural design. Function of the exercise is set as small family independent living space, attached with a small cultural saloon. Family member includes a young couple (with background of design industry) and a child (7~12 years old). New building area is about 160~200 m^2 (±5%), building height <8 m (total height of internal usable space, excluding parapet, no underground space), in which the living space shall include living room, dining room, master bedroom, children's bedroom, multifunctional room (including working space and guest bed), kitchen, dining room, toilet (1~2) and necessary storage space and shall meet the function of cultural saloon, including reading area, cultural display and saloon space, storage area, etc. Cultural saloon shall be rationally combined with residential space.

(3) Streamline and access: Internal functional space of building shall not only be mutually communicated and contacted by rational horizontal and vertical traffic, but also shall consider rational connection with surrounding environmental condition, such as continuity of street interface, selection and setback of access, influence of exterior wall of surrounding building (including exterior window) on newly built building, spacing and sight interference between buildings, rational use of daylight, etc. Internal stair of new building must meet current code (step width 25~30 cm, step height 15~20 cm, continuous step quantity is not more than 18, etc.).

(4) Dimension and sensing: Internal space of building is used by human, therefore, dimension of functional space in the building must take human body as basic reference and consideration, and confirm rational building space dimension according to various human behaviors and activity modes. Relationship between space composition and space experience shall be understood through graphic expression in space form treatment, mainly including senses of dimension and enclosure.

(5) Structure and construction: Select rational and effective supporting system and wrapping system according to specific space characteristics, complete space limit and visual effect through a series of comprehensive material, structure and construction design.

Teaching Arrangement

The design courses last for 7 weeks (May 4, 2018–June 22, 2018)

First week: Survey site, group make 1∶100 site model (seat 60 cm×60 cm×5 cm), use working model to conceive preliminary plan, research relationship and effect of site and interface through picture collage, frontage perspective.

Second week: Use 1∶50 drawn plan, facade and section drawing to comprehensively research function and space, streamline and dimension.

Third week: Confirm basic design plan, promote section, facade design, use working model to assist (centralized drawing comment).

Fourth week: Deepen design plan and refine design detail, model and research internal space effect.

Fifth week: Detail 1∶50 drawings, make picture collage effect drawings (centralized drawing comment).

Sixth week: Make 1∶50 section perspective and analytic drawings, make 1∶50 model (centralized drawing comment).

Seventh week: Arrange drawing, typeset and complete course defence.

Result Requirement

3 A1 drawings, 1 paper expressed model (scale 1∶50), several working and research models. Drawing content shall include:

(1) General plan (1∶200), plan of levels, longitudinal and transversal section drawings and main facade (1∶50), 1 internal space organization and construction expression section perspective (1∶50).

(2) Design specification and main technical and economic index (land area, building area, plot ratio, building density).

(3) Analytic drawings that expresses design intention and design process (block generation, function analysis, circulation analysis, structure system, etc.).

(4) Paper model photos and computer effect drawings, photo collage, etc.

建筑中的各功能空间的尺度,都必须以人体作为基本的参照和考量,并结合人体的各种行为活动方式,来确定合理的建筑空间尺寸。
Dimension of functional space in the building must take human body as basic reference and consideration, and confirm rational building space dimension according to various human behaviors and activity modes.

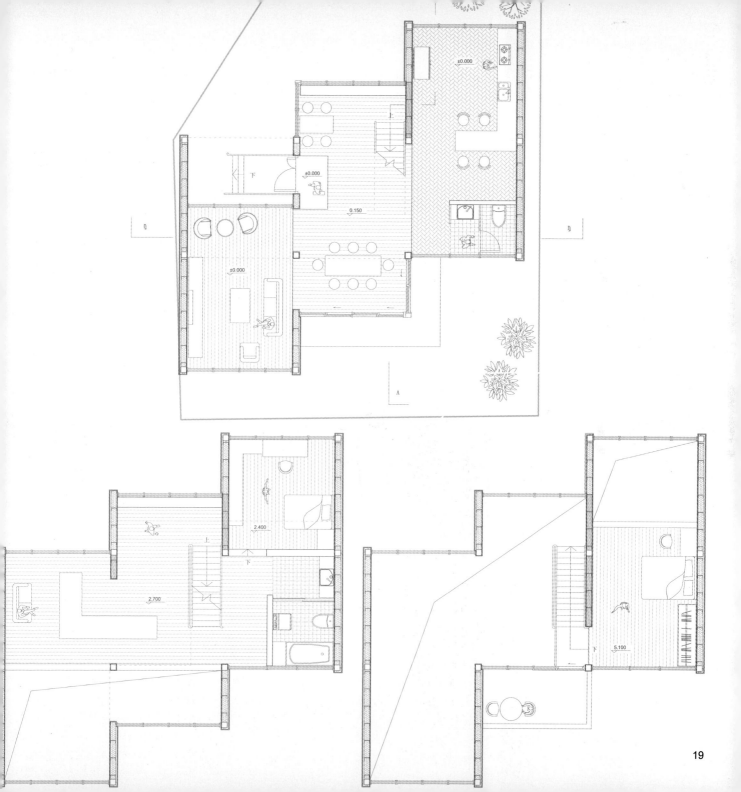

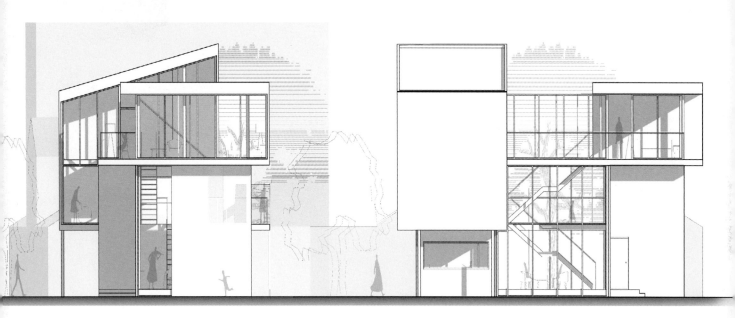
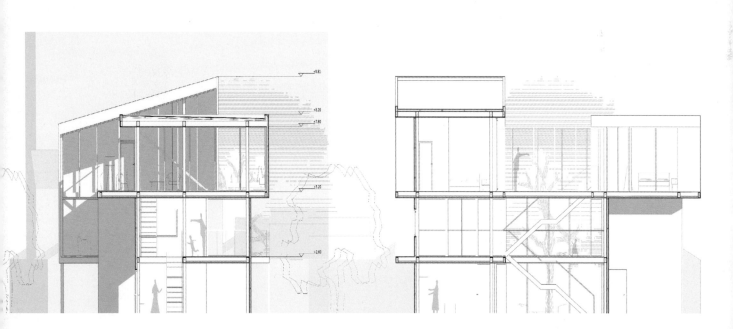

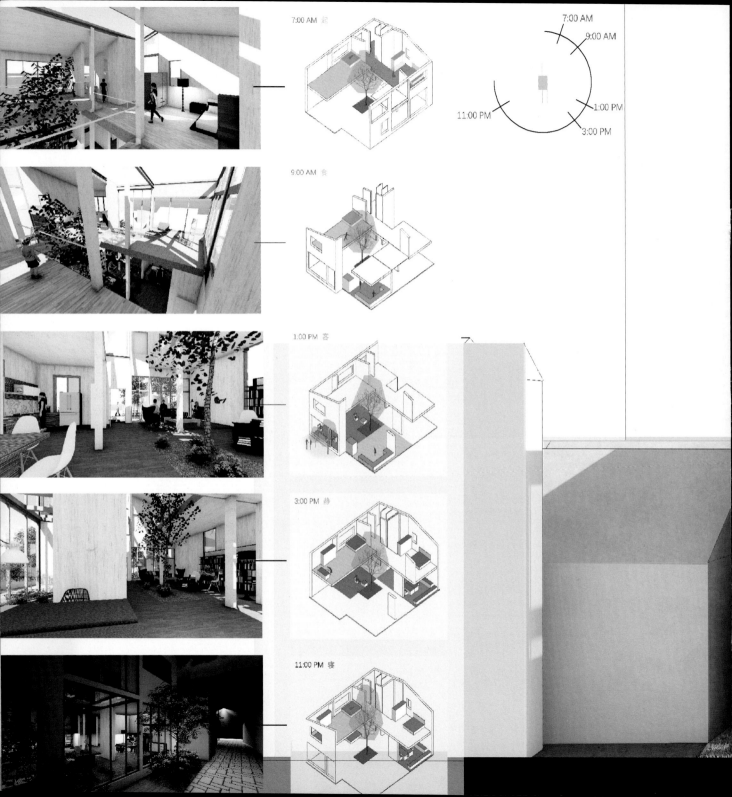

混凝土板
防水卷材
保温层
钢筋混凝土层
压型钢板
钢梁
刨花板
涂层

外墙坚挂板
防水卷材
刨花板
横龙骨
工字钢（保温层）
刨花板
涂层

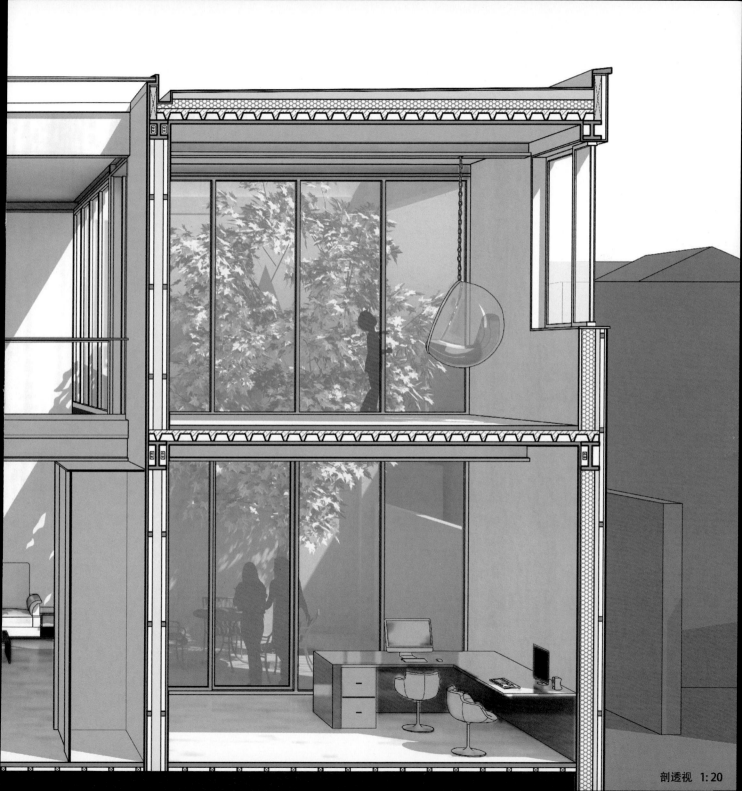

剖透视 1:20

乡村小型家庭旅馆扩建
EXTENSION OF A COUNTRY HOUSE AS HOLIDAY INN

建筑设计（三）ARCHITECTURAL DESIGN 3

周凌 童滋雨 窦平平

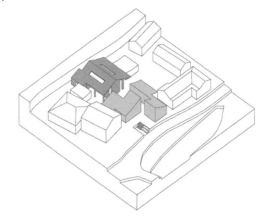

教学目标
此课程训练解决两个基本问题：一是房屋结构、材料、构造等建造问题；二是基本起居、居住功能的平面功能排布问题。通过这一建筑设计课程的训练，使学生在学习设计的初始阶段就知道房子如何造起来，深入认识形成建筑的基本条件：结构、材料、构造原理及其应用方法，同时课程也面对地形、朝向、功能问题。训练核心是结构、材料、构造、基本功能，强化认识建筑结构、建筑构件、建筑围护等实体要素。

教学内容
观音殿村位于南京市江宁区秣陵街道，由于城乡统筹发展与乡村治理的需要，需要对村内现有房屋进行改造，部分房屋改造为乡村公共配套服务建筑，部分改造为对外服务与经营用房，部分改造为小型家庭旅馆。每个基地保留1~2栋老房子，改造为客房。另外在院子内进行加建设计，加建部分作为作坊、展示、客厅、餐厅等公共功能。

规划要求
建筑层数1~2层，建筑限高：檐口高度不超过7.5m，总高不超过9m。平顶坡顶不限。要求充分考虑材料建造与实施的可能性。

改造部分客房面积：单间20~30m²，套间30~45m²。加建公共部分面积约100~300m²。

材料建造
材料结构有预先准备的材料清单和结构选型。围合与覆盖材料可以选择砖、瓦、木、石、土、金属、玻璃、塑料等。主要结构材料必须在指定材料中选择，其他材料和辅材自定。

Teaching Objective
This course training solves two basic issues: one is the construction issue as housing structure, material, construction, etc.; another is the plan function arrangement issue as basic living, living function. Through the training of this architectural design course, students know how to build a house at initial stage of design study, deeply realize the basic condition for forming architecture: structure, material, construction principle and application method, at the same time, the course also aims at landform, orientation, function issue. The training core is structure, material, construction, basic function, reinforcing cognition of architectural structure, architectural component, architectural enclosure, etc..

Teaching Content
Guanyindian Village is located in Moling Street, Jiangning District, Nanjing. Due to the requirement of urban-rural integration development and countryside control, it is required to reform existing houses in the village. Some houses are reformed into rural public supporting service building. Some houses are reformed into external service and operation houses. Some houses are reformed into small family inn. 1~2 old houses are reserved for each base, reformed into guestroom. In addition, additional construction is designed in the courtyard, the increased part is used as workshop, display space living room, dining room, etc..

Planning Requirement
Building has 1~2 floors, building height is limited; cornice height is not more than 7.5m, total height is not more than 9m. Flattop or slope crest is not limited. Possibility of material construction and implementation shall be fully considered.

Reform part of guestroom area; single room 20~30m², suite 30~45m². Increased public area is about 100~300m².

Material Building
Material structure has a pre-prepared material list and a structure type selection. Brick, tile, wood, stone, earth, metal, glass, plastic, etc. can be selected as enclosing and covering material. Main structure material must be selected from the designated material, other material and auxiliary material are independently confirmed.

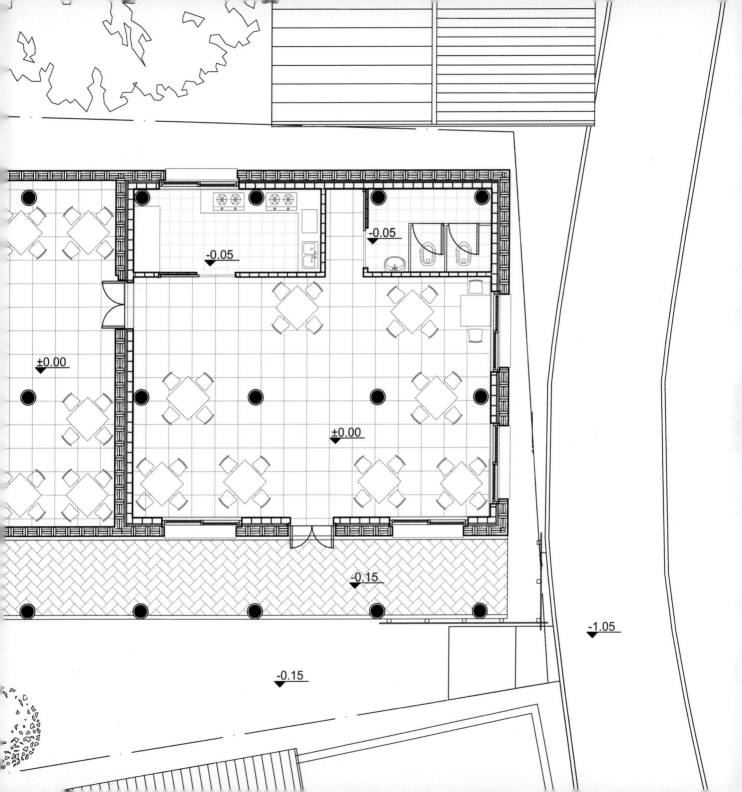

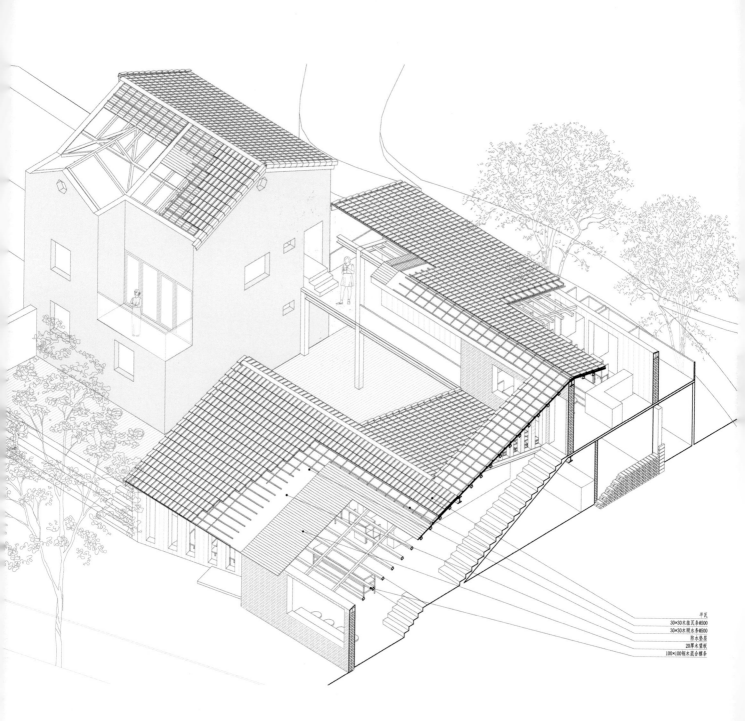

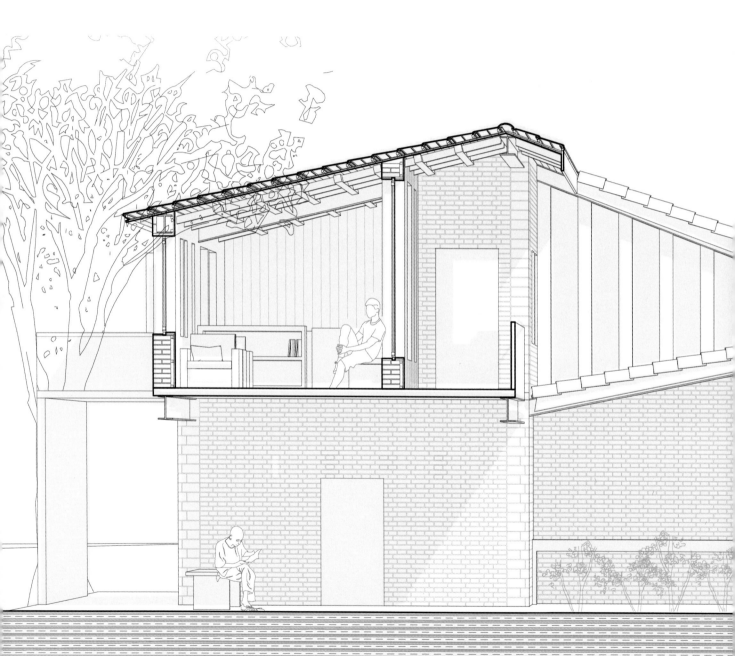

建筑设计（五+六）ARCHITECTURAL DESIGN 5 & 6

城市建筑：大型公共建筑设计
URBAN ARCHITECTURE: COMPLEX BUILDING

华晓宁 钟华颖 王铠

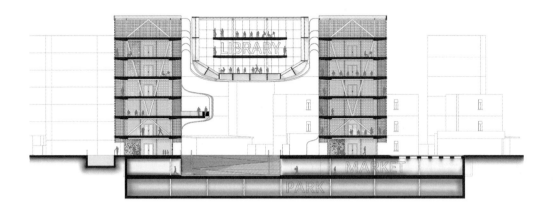

近年来对"日常性"的关注与其说是一种新的风潮，毋宁说是对建筑学真正本体的回归。在此，所谓"本体"不仅仅是"自身"（亦即建立在物质性基础上的内在逻辑性或自明性），更是一种"本原"，是建筑学得以缘起和确立自身价值的根基。

城市之于建筑，无疑是一种背景结构或基质式的存在，它对所处其中的建筑的意义、影响、限定大概可以从两个方面来理解，亦即物质空间层面和日常生活层面。前者往往有关于几何学，后者则更适合以现象学视角为切入点。城市在物质空间和日常生活两方面的复杂性，成为"城市建筑"最根本的缘起和约束。"城市建筑"不仅仅需要物质空间层面的整合和适配，更重要的是物质空间所链接和承载的城市日常生活。

南京大学建筑系本科三年级下学期建筑设计课程聚焦于"城市建筑"这一主题已有多年。尽管以往在教学中对城市建筑的功能、空间、动线、形体组织训练做了许多探讨，还进一步引入了"实与空""内与外""层与流""轴与界"等几组关键词来启发学生关注和理解建筑与城市空间的整合，但始终存在的问题是过于强调操作性训练，过于强调物质空间本体，对于物质空间所承载、激发和伺服的真实城市生活以及城市场址上具象的使用者关注不够，造成了学生难以真正"沉浸"到真实的城市日常状态中，难以真正关注具体的市民和鲜活的日常生活，从而造成"城市建筑"中城市与建筑的貌合神离。另一方面，学生的年龄和阅历决定了他们不太可能对城市具有丰富而深入的经验和理解，需要一些工具帮助他们深入城市日常生活。

因此，在2018年度的课程教学中，教学组引入"城市针灸"和"城市触媒"这两个新的教学关键词，引导学生关注城市"日常性"，并将其作为"城市建筑"设计概念和策略的源泉和出发点。

"城市针灸"最早进入国内建筑学界视野，是在2004年南京大学建筑系主办的"结构·肌理·地形学"国际学术研讨会上。当前国内方兴未艾的"微更新"则是这一策略在当前存量更新背景下的在地实践。"城市针灸"有两个最重要的特征。一是"准"，必须精准地发现城市日常生活中切实存在的真实问题，并提出切实可行之有效的解决策略，而这必然建立在对城市"日常性"许尽深入研究、分析的基础上。二是"微"，必须以尽可能谨慎、小尺度的操作来解决问题，以最小的代价和成本获得尽可能好的收益，并避免惯用的"大手笔"给城市带来副作用。

"（城市触媒）策略性地引进新元素以复苏城市中心现有的元素且不需彻底地改变它们，而且当触媒激起这样的新生命时，它也影响了相继引进之都市元素的形式、特色与品质。"广义上说"城市触媒"并不一定是建筑，但"城市建筑"毫无疑问应成为城市进程中的一种重要活化"触媒"，促使原有城市环境加速更新，促使原有城市结构持续演进。这就需要将建筑的物质空间系统与丰富的日常生活紧密关联起来，通过空间的多样化计划性安排，为物质空间带来活力。

"城市针灸"和"城市触媒"这两个关键词的共同特征是对于"日常性"的重视。它们既以城市"日常性"为源头，又以催生新的、更为丰富的"日常性"为导向和指归。另一方面，尽管"物质性"和"形态"依然是最为重要的操作性工具，但"效能"无疑才是最终的评价标准。

本年度的设计课题依然是在南京五塘新村中置入一个新的社区中心。五塘新村社区位于主城边缘，始建于1980—1990年代，是一个当代中国城市中非常典型、大量可见的老旧高密度居民小区，人口密集，成分复杂，设施老旧，管理松懈，空间环境杂乱却又生机勃勃。理解此类城市环境，"日常性"无疑是最重要的切入点。而从

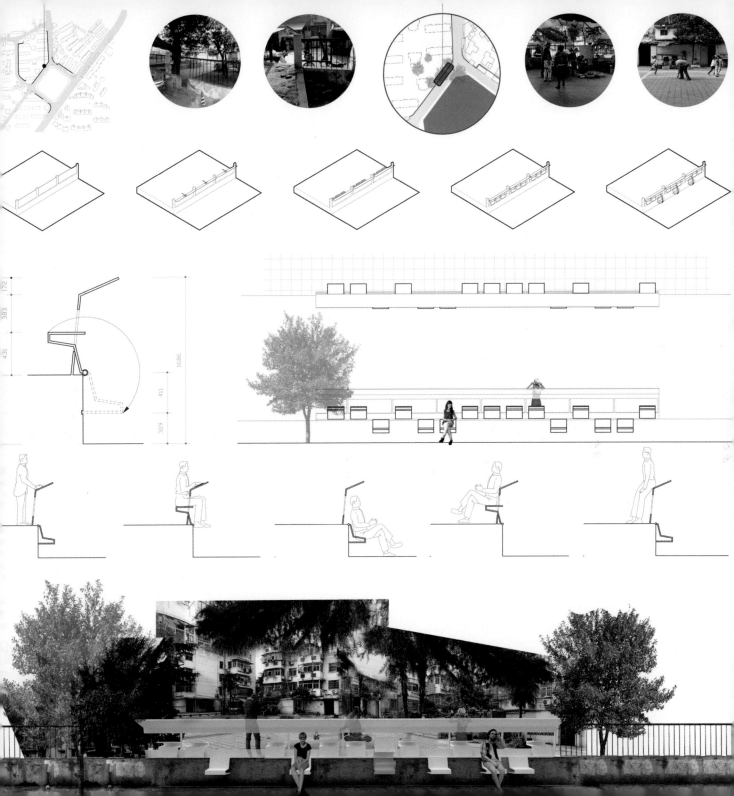

另一个角度来看,此类城市场址,正是学习以"日常性"指导"城市建筑"设计的最佳标本。

16周的教学周期被分为两个教学阶段。

在第一阶段,学生们被要求仔细深入地实地调研整个社区,运用各种手段记录、分析社区物质空间系统的形态以及社区居民的日常生活状态,发现社区生活中的真实问题,了解社区居民的真实需求。教学组尤其强调照片、影像的工具性作用,此外也鼓励学生自行拓展其他多种多样的城市环境和城市日常性分析工具和途径。在教师的鼓励和启发引导下,学生表现出了极大的主动性和创造性,例如有学生就从社区小学网站上发表的学生作文中提取了居民在社区外部空间环境中的日常行为、感受和评价。

在调研、记录、分析的基础上,学生需要展开一次"城市针灸":在社区中心用地周边选择一处存在问题和需求的外部空间进行"微更新",用尽可能小的介入改变空间环境,解决真实问题,切实满足居民需求。"微更新"操作被要求不能仅仅停在"满足需求"的层面,还必须进一步以此为契机激发一系列衍生的市民活动和人际交往。"城市针灸"成为整个课题训练的预热和触发点,从某种意义上说,它本身也成为一种教学的"触媒"。

第一次阶段性评图被命名为"五塘医案"。这一名称来自教学过程中学生工作内容和方法的类比:对场地的调研分析过程对应于中医的"望闻问切",设计问题、目标和策略的提出类似于中医开方,而"社区微更新"则是典型的"针灸"操作。为了进一步激发学生的主动性和积极性,教学组在阶段性评图的同时举办了名为《五塘新村的日常》的影像展。

教学的第二阶段,则要求学生进一步将目光聚焦到社区中心的设计,它被要求设计成为一个"触媒",以物质空间激发城市生活,催化城市活力,诱发多种多样的城市"日常性"。为此,本学期延续了上一学年"开放式任务书"的做法,限定的必需功能空间仅占总建筑面积的1/4,剩余3/4空间的性质、类型和计划需要学生通过研究来自行确定。

在设计成果方面,除了常规的技术图纸之外,特别要求学生完成"空间叙事"的表达:以一系列小透视表达新建筑介入城市场址后产生的一系列新的建筑与城市空间片段。这些空间片段场景中必须加载未来可能持续发生或在不同场合下发生的市民使用行为,它们是由新建筑介入而引发的"链式反应"的一部分,是新的城市"日常性"。

此外,学生需要考虑将前一阶段完成的"城市针灸"(微更新)与作为"城市触媒"的社区中心相整合:或是将某些微更新策略进一步应用到社区中心的设计中,或是令微更新的成果与社区中心共同作用,形成具有活力的城市场所。

在教学中强化对"日常性"的重视,在本质上是引导学生向建筑学"本原"和"初心"的回归。这种具象的观察力、独立的思考力对学生未来的职业生涯是极为重要的。

In recent years, the concern of everydayness is rather a return to the noumenon of architecture than a new tide. The so-called "noumenon" is not only "self" (internal logicor or self-evidence established on material basis), but also a "principle", the foundation for origin and value of architecture.

City is no doubt an existence of background structure or matrix to architecture, its meaning, influence, limit on architecture in it can be understood from two aspects, physical space and everyday life. The former is often associated with geometry, and the latter is more suitable to start with the view of phenomenology. The complexity of city in physical space and everyday life becomes the fundamental origin and restriction of "urban architecture". "Urban architecture" does not only require the integration and adaptation of physical space, but also require the urban everyday life linked and carried by the physical space.

Architecture Design in the second semester of the third year undergraduate in Department of Architecture of Nanjing University has focuses on "urban architecture" for years. Although many discussions have been made on the function, space, motion line, shape organization training in the teaching, several groups of keywords as solid & void, inner & exterior, layer & circulation, axis & edge are introduced to inspire students to concern and understand the integration of architecture and urban space, but the problem is operational training and physical space body are over-emphasized, the concern of real urban life carried, inspired and served by physical space and actual user on the urban site is insufficient, students cannot be really "immersed" in real urban daily life, cannot really concern specific citizen and vivid everyday life, as a result, city and architecture in "urban architecture" are seemingly in harmony but actually at variance. On the other hand, age and experience of students restrict them from having rich and profound experience and understanding of city, they need some tools to go deep into the urban everyday life.

Therefore, in the course teaching in 2018, the teaching group introduces two new keywords, "urban acupuncture" and "urban catalyst", to guide students to concern urban "everydayness" and take it as the source and start of "urban architecture" design concept and policy.

"Urban acupuncture" came into sight of domestic architecture at "Structure, Fabric and Topography" international symposium seminar sponsored by Department of Architecture of Nanjing University in 2004. "Micro updating" in ascendant is the practice of this policy under the background of stock updating. "Urban acupuncture"

has the most important two characteristics. One is "precise", the real problem in urban everyday life must be precisely found, and effective solution must be proposed, which is established on the basis of detailed research, analysis of urban "everydayness". Another is "micro", problem must be solved as cautiously as much in small size, so as to obtain better income at the minimum cost, and avoid side effect on the city caused by "overwhelming strategy".

(Urban catalyst) strategically introduces new element to revive existing element in urban center without completely changing them. When catalyst evokes such new life, it also affects the form, characteristics and quality of subsequently introduced "urban element." In a broad sense, "urban catalyst" may not be architecture, however, "urban architecture" is no doubt an important activated "catalyst" in urban progress, promoting updating of original urban environment and continuous evolvement of original urban structure. It requires closely associating physical space system of architecture with rich everyday life to bring vigor to physical space through diversified planned arrangement of space.

Both "urban acupuncture" and "urban catalyst" attach importance to "everydayness". They originated from urban "everydayness" and catalyze new and richer "everydayness". On the other hand, although "physicality" and "form" are still the most important operational tool, performance is no doubt the final evaluation standard.

The design task in this year is still building a new community center in Nanjing Wutang New Village. Wutang New Village Community is located on the edge of main urban area, established in 1980s and 1990s, as a very typical and common old high density resident community in modern China, where the population is dense, the composition is complex, the facility is old, the management is loose, space environment is disordered but vigorous. To understand such an urban environment, "everydayness" is no doubt the most important entry point. From another perspective, such urban site is just the best sample to study "urban architecture" design guided by "everydayness".

The 16-week teaching is divided into two phases.

In the first phase, students are required to carefully have field survey of the whole community, apply various methods to record, analyze the form of community physical space system and everyday life of community resident, find out the real problem in community life, and understand the real demand of community resident. The teaching group especially emphasizes the tool function of picture, video, and encourages students to independently expend other tool and approach to analyze urban environment and urban everydayness. Under the encourage and inspiration from teacher, students show great initiatives and creativity, some student extracts daily behavior, feeling and evaluation of resident in space environment out of community from the writing issued on community primary school website.

On the basis of survey, record, analysis, students shall conduct "urban acupuncture": choose an exterior space with problem and demand surrounding the land of community center for "micro updating", intervene to change space environment as least as possible, solve real problem, effectively meet the resident requirement. "Micro updating" cannot only stay on the level of "meeting requirement", but also must inspire a series of derivative citizen activities and interpersonal communication. "Urban acupuncture" becomes the warm-up and trigger of the whole task training, in some sense, it becomes a kind of teaching "catalyst".

The first stage design comment is named as Wutang Medical Case. This name is from the analogy of job content and method of students in the teaching: the field survey analysis process corresponds with "four ways of diagnosis" in traditional Chinese medicine, proposal of design problem, objective and strategy is similar to prescription of traditional Chinese medicine, while "community micro updating" is a typical "acupuncture". To further inspire the initiatives and activeness of students, the teaching group also organizes photo show named Everydayness of Wutang New Village in the stage design comment.

In the second phase, students are required to focus on the design of community center. It is required to be designed into a "catalyst", inspire urban life by physical space, catalyze urban vigor, and induce various urban "everydayness". Therefore, the method of "open assignment" in last year is continued in this semester, the restricted required function space only accounts for 1/4 of total floor area, nature, type and program of the rest 3/4 space shall be independently confirmed by students through research.

On design achievement, besides routine technical drawing, students are especially required to complete "space narration": use a series of small perspective to manifest a series of new segments of architecture and urban space after new architecture intervenes in urban site. These space segments must be loaded with citizen use behaviors that may continuously occur in different occasions in the future, they are a part of "chain reaction" caused by intervention of new architecture, as new urban "everydayness".

In addition, students shall consider integrating "urban acupuncture" (micro updating) completed in the previous phase with community center as "urban catalyst": or further applying some micro updating strategy in the design of community center, or functioning the result of micro updating with community center together, to form vigorous urban site.

Strengthening emphasis on "everydayness" in teaching is to essentially guide students to return to "principle" and "original will". Such figurative outsight, independent thinking are extremely important to the future career of students.

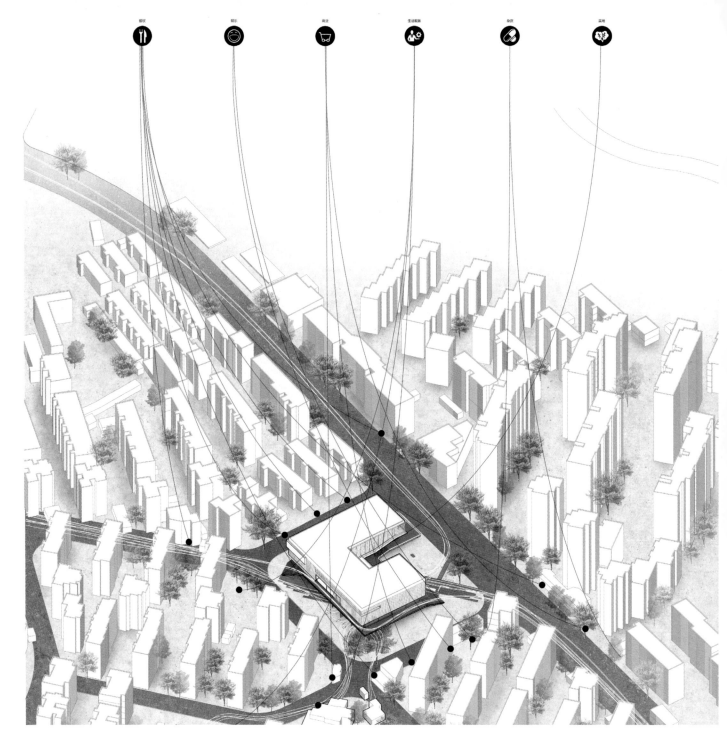

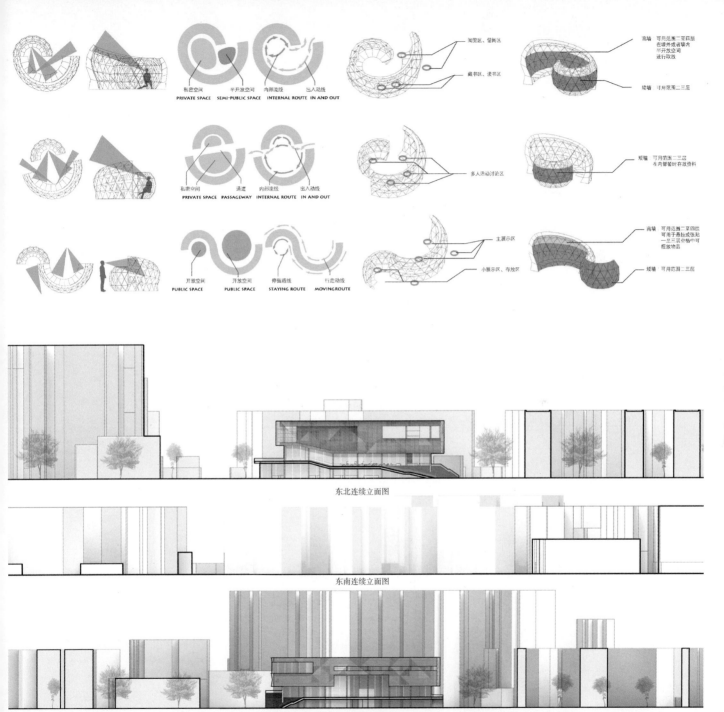

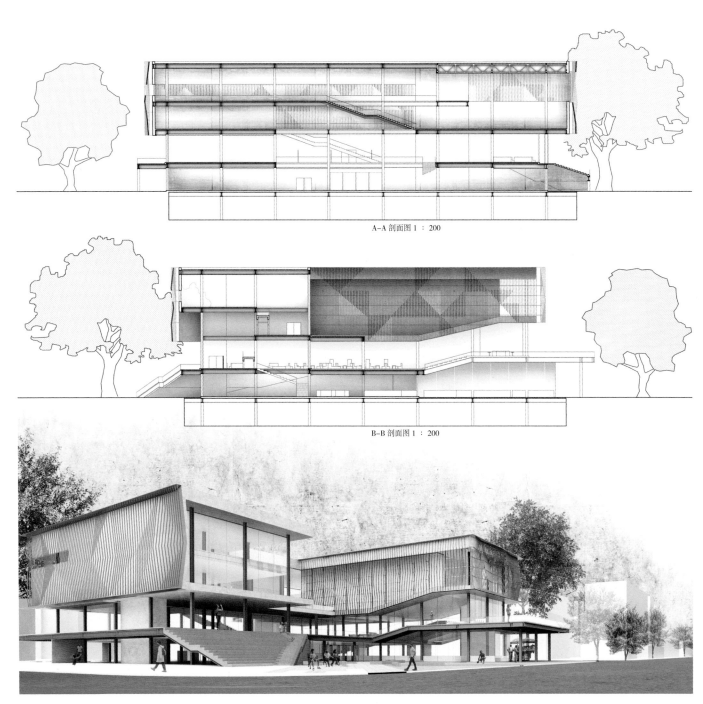

A-A 剖面图 1：200

B-B 剖面图 1：200

通高的展览空间与二层的室外平台连通，室内室外的视线相互交流。展览内容被室内和室外同时感知。

展览空间伴随中庭的交通空间布置，使人产生连续的观看体验，人们在不知不觉中已经走进了社区中心的平台之上。

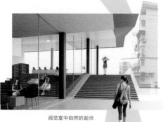

阅览室中自然的起伏形成了不同种类的阅读空间，外面让通过的居民也能看到室内的活动情况，令人产生走进去加入的想法。

平台围合形成的庭院以及平台外沿形成的灰空间，中人们正在休息交谈。平台延伸到地面的台阶使人们随时可以走入建筑中。

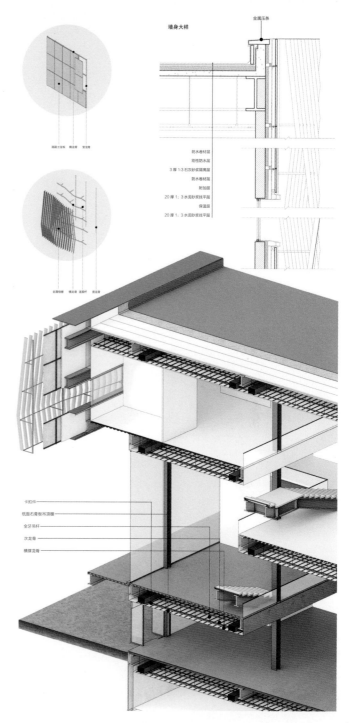

45

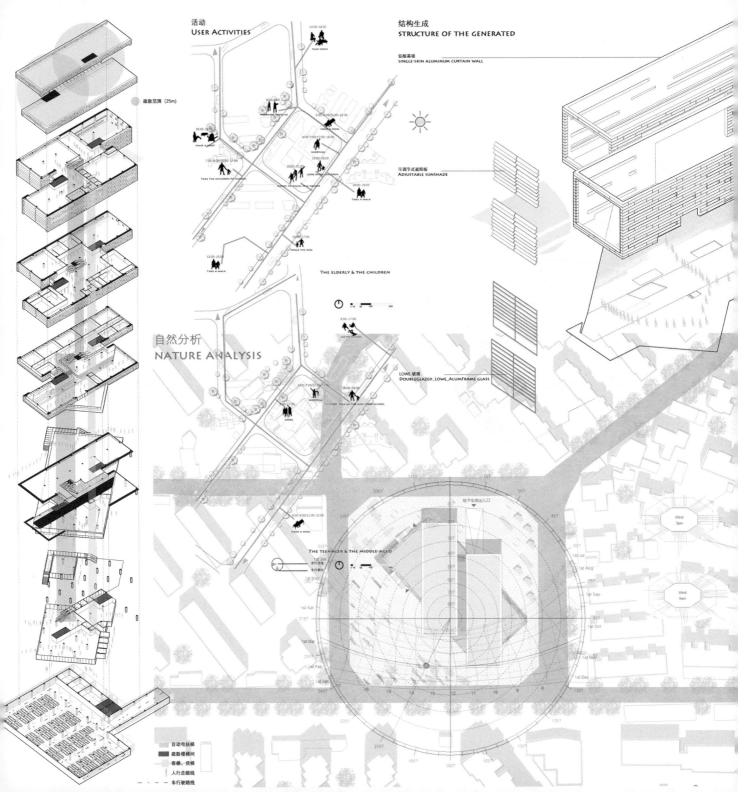

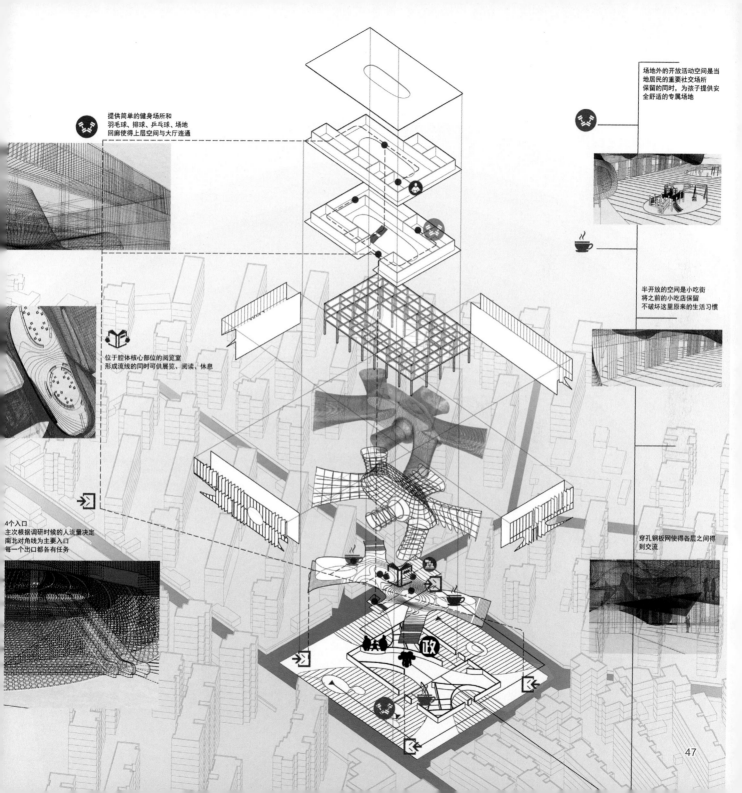

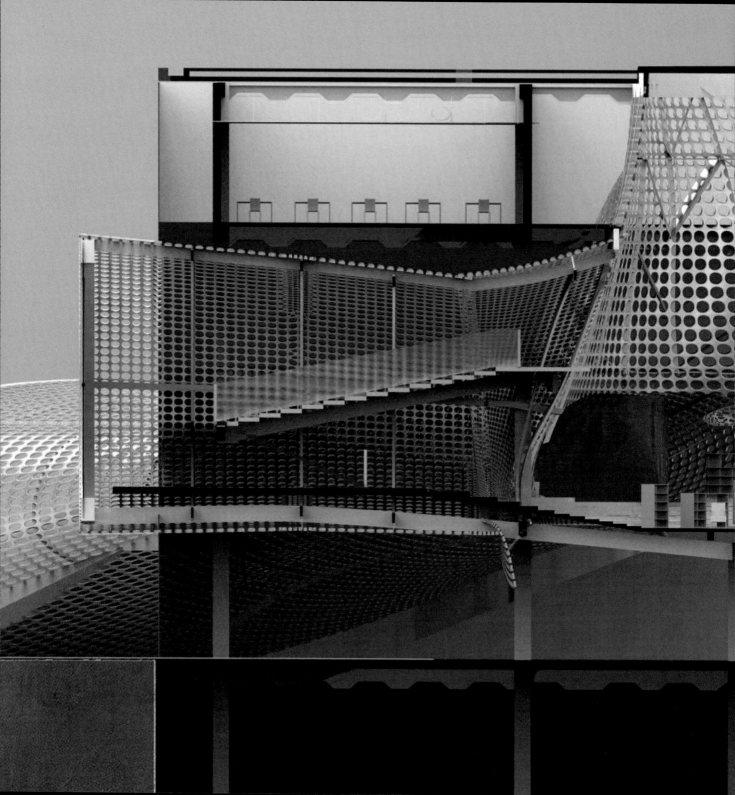

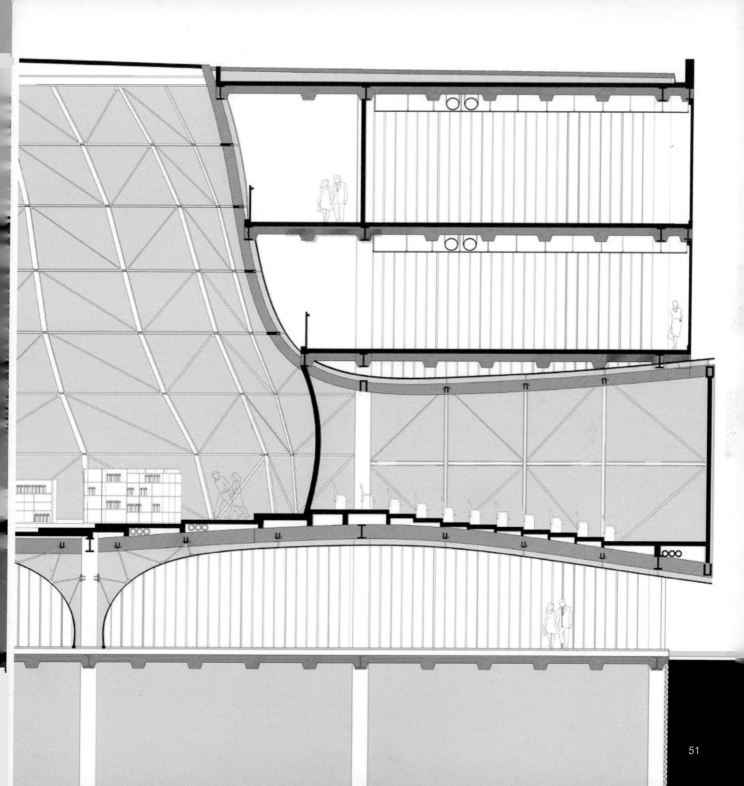

本科毕业设计 GRADUATION PROJECT

互动建筑原型数字化设计与建造
INTERACTIVE ARCHITECTURAL PROTOTYPE DIGITAL DESIGN AND CONSTRUCTION

吉国华 孙彤

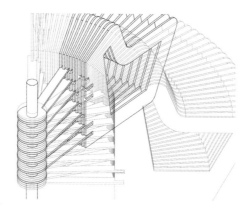

互动建筑作为智能建筑在建筑本体上的深化，在增强建筑的物理环境品质、提高建筑空间的吸引力与体验感方面有优越性，其作为一项实践性极强的研究探索目标是达成建筑空间与人的交互性。互动建筑的研究对象是建筑本身的支撑、分隔与围护结构及灯光照明系统等与人关联的终端系统智能动态性，如自动感应室外自然光强弱而进行建筑立面遮阳几何形态调整的建筑立面结构，以及可以感应人的走动停留而产生形态变化的建筑幕墙系统，甚至可以自主进行动态调整改变空间几何形态的轻型支撑结构。

自2012年起，南京大学建筑学的本科毕业设计一直在进行探索数字化设计与建造的教学，并开展了一系列专题化的教学实践。在此基础上，2018年南京大学本科毕业设计的数字化教学尝试引入互动建筑技术。本次课程在互动建筑原型数字化设计与建造大的方向下，主要分为"动态结构"与"光影界面"两个专题。本次教学将面向原型设计与装配的实验教学重点落在机械传动设计与程序算法之上，展现了在继承传统的面向设计案例的建构教学基础上，对数字技术的融合与运用。

本课程第一阶段的主要任务是理解由单元到整体的互动建筑设计概念。电子媒体图像是由像素点所构成的，如果把图像理解为整体，像素点就是构成整体的单元。在抽象概念上整体具有独特性，它们源自不同层级组织和整合之间连续的相互作用。因此整体并不是简单的单元物理累加，整体包含了单元行为之间的关系与成组控制，单元需要通过编程进行统一整合。在现实意义上，从单元入手不仅有助于提高整体的稳定性，而且有助于构架循序渐进的学习体系，从而便于从克服简单的问题入手逐步迈向系统问题的梳理与解决。

本课程第二阶段的训练核心是互动技术涉及的控制器、输出和输入三个主要部分的技术性测试。首先对单片机的学习选择较为容易掌握的arduino系列芯片，学生首先要学习可以读取和输出开关信号（对应数值1和0）电压的数字引脚可以读取和输出模拟信号（对应数值0~255）电压的模拟引脚，并且通过对LED灯泡灯效的测试掌握对数字和模拟引脚的使用。然后，进入动力输出设备的学习测试阶段，该阶段主要选择直流电机、舵机或步进电机作为输出设备，用数字引脚或模拟引脚分别实现开关、角度和速度的控制。最后，进入传感器的学习的测试阶段，该阶段主要选择光敏电阻和超声波传感器作为输入设备，光敏电阻是常用的非接触式开关输入设备，其优点在于不需要时间函数，因而能保持连续稳定的工作状态，缺点是受环境光条件影响较大且探测距离较近。超声波传感器的使用需要借助编程调用时间函数计算前次超声与返回超声的时间差来计算距离，因此超过一定工作时间需要重置时间函数，但其具有探测区间较大的优点。

本课程第三阶段的训练核心是原型的建构与面向最终建造的状态控制设计。学生首先需要完成运行稳定的实物原型，解决实物原型所遇到的传动与摩擦问题，并在此基础上思考最终作品在不同的环境状态与使用情形下形成的整体可能对人的活动产生的不同状态的响应模式。在完成的实物原型中，要求模型能够切实与稳定的物理动态效果达成互动效果。在此基础上，对实体建筑愿景建造场景中需要注意与解决的构造问题做出分析。

通过本课程的教学，也暴露了一些问题。首先，建筑学学生知识背景偏重设计概念与绘图，而对基本弱电电路与计算机编程技术的缺乏造成这个区域成为学生的知识盲区，设计概念难以落地。通过设计案例教学的方式，指导教师给学生提供亲身尝试数字技术的机会与技术帮助，鼓励对数字技术的自主学习与运用，使得数字技术成为一种工具帮助设计的深入推进。其次，教学过程中对于互动数字技术的运用还只是停留在概念测试阶段，互动模型与可以推向市场的建筑智能产品相比无论在完善度还是在可靠度方面都存在较大的差距。在今后的互动建筑教学中将尝试与电气设备制造商的合作，以提升互动建筑环境作品的完成度。

As the detailing of intelligent architecture, interactive architecture is superior in strengthening physical environment quality of architecture, improving attraction and experience of architectural space, and it aims to realize interaction between architectural space and human as a research with extremely strong practice. The research object of interactive architecture is the intelligent dynamic nature of human related terminal system as support, partition and enclosing structure of architecture, such as facade structure that adjusts facade sunshade geometric form according to automatic sense of outdoor natural light, and curtain wall system that has form change by sensing walking and staying, and even the light supporting structure that can automatically have dynamic adjustment and change of space geometric form. Since 2012, the undergraduate diploma project in Department of Architecture of Nanjing University has been exploring the teaching research of digital design and construction, and conducted a series of specialized teaching practice. On this basis,

the digital teaching for undergraduate diploma project in Nanjing University in 2018 attempts to introduce interactive architecture technology. Under the direction of interactive architecture prototype digital design and construction, this course is mainly divided into "dynamic structure" and "light and shadow interface". The experiment teaching of prototype design and assembly focuses on mechanical drive design and programmed algorithm, and shows the integration and application of digital technology on the basis of inheriting traditional tectonic teaching which was aimed at architecture projects.

The main task in the first phase of this course is to understand the interactive architectural design concept from unit to whole. Electronic media image is composed by pixel point, if image is a whole, pixel point will be the unit to compose the whole. In abstract concept, whole is unique, from the continuous mutual function among organizations and integrations of different levels. Therefore, the whole is not only simple physical superposition, but includes the relation and group control among unit behaviors, and unit is integrated through programming. In realistic sense, starting with unit is not only helpful to improve the stability of the whole, but also helpful to construct progressive learning system, so as to gradually transfer from overcoming simple problem to rationalizing and solving system problem.

The training core of the second phase of this course is technical test of three main parts of interactive technology, that is, controller, output and input. First, arduino series chip easy to master is selected for study singlechip, students study digital pin that can read and output switch signal (correspond with value 1 and 0) voltage and analog pin that can read and output analog signal (correspond with value 0~255) voltage first, and master the use of digital and analog pin through testing effect of LED bulb. Then, the learning test of power output equipment starts, at this stage, students mainly choose DC motor, steering engine or stepping motor as output equipment, use digital pin or analog pin to respectively realize control of switch, angle and speed. Finally, the learning test of sensor starts, at this stage, students mainly choose photoconductive resistance and ultrasonic sensor as input equipment. Photoconductive resistance is common non-contact switch input equipment with advantage of not requiring time function and maintaining continuous and stable working status, disadvantage of being affected by ambient light condition and limited detection distance. For use of ultrasonic sensor, time function shall be called through programming to calculate the time difference of previous ultrasonic and returned ultrasonic to calculate distance, therefore, time function shall be reset when exceeding certain working time, however, its detection range is large.

The training core in the third phase of this course is construction of prototype and final construction status control design. Students shall complete stably operated physical prototype, solve the drive and friction problem of physical prototype, and think the whole formed by final work may product response mode in different states on human activity in different environmental states and use circumstances on this basis. Among the completed physical prototype, the model shall effectively meet the interaction effect with stable physical dynamic effect. On this basis, construction problem that shall be concerned and solved in the physical architecture construction scene shall be analyzed.

Some problems are also found in the teaching of this course. Firstly, students of architecture lay particular stress on design concept and drawing in knowledge background and lack of basic weak current circuit and computer programming technology, which causes knowledge dead zone and makes it difficult to realize design concept. In the method of design case teaching, the tutor provides students with opportunity of personally trying digital technology and technical aid, encourages students to independently study and apply digital technology, and makes digital technology help deeply promote design as a tool. Secondly, the application of interactive digital technology is only at the conceptual test stage in the teaching, compared to the intelligent architecture product that can be promoted to market, the interaction model is still inferior in both degree of completeness and reliability. We will try to cooperate with electric equipment manufacturer in the future interactive architecture teaching and improve the degree of completeness of interactive architecture environment work.

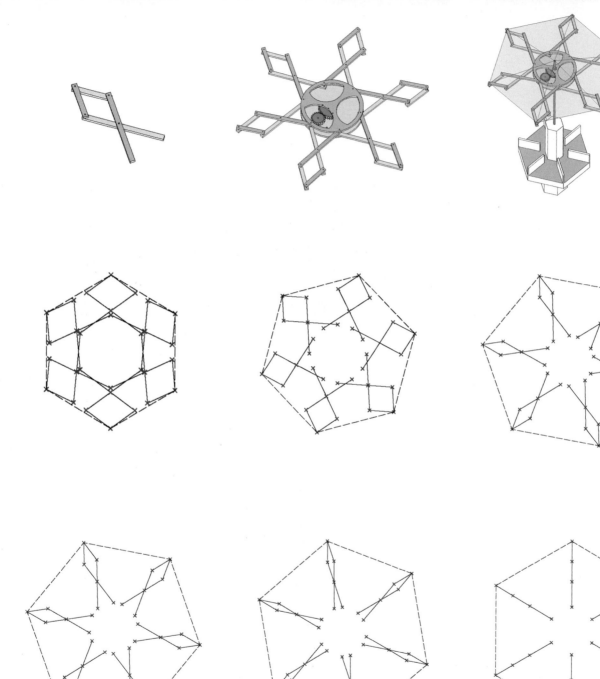

单元拼接
Unit merging

角度变化
Angle variation

步骤1：切割出设计好的木板作为模型主体结构的材料
Step 1: Cut the designed wood board as the material of the main structure of the model

步骤2：将伸展结构的第一部分与两个圆盘分别固定
Step 2: Fix the first part of the stretching structure with two disks separately

步骤3：将大齿轮固定在上方圆盘上
Step 3: Fix the big gear on the upper disc

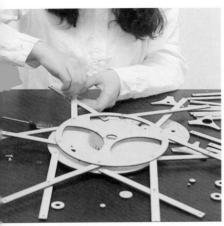

步骤4：上下两个圆盘上伸展结构的第一部分交叉固定
Step 4: Cross-fix the first part of the stretching structure on the discs

步骤5：将伸展结构的第二部分与第一部分固定
Step 5: Fix the second part of the stretching structure to the first part

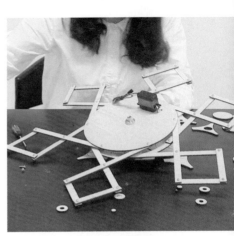

步骤6：将舵机固定在下方圆盘上
Step 6: Fix the steering engine on the lower disc

步骤7：将下方圆盘的中心固定在支撑机构上
Step 7: Fix the center of the lower disc on the supporting mechanism

步骤8：裁制出适合形状和大小的半透明布料
Step 8: Sew translucent fabric suitable for shape and size

步骤9：将布料缝制在伸展结构之间的弹性材料上
Step 9: Sew the tailored fabric onto elastic materials between stretching structures

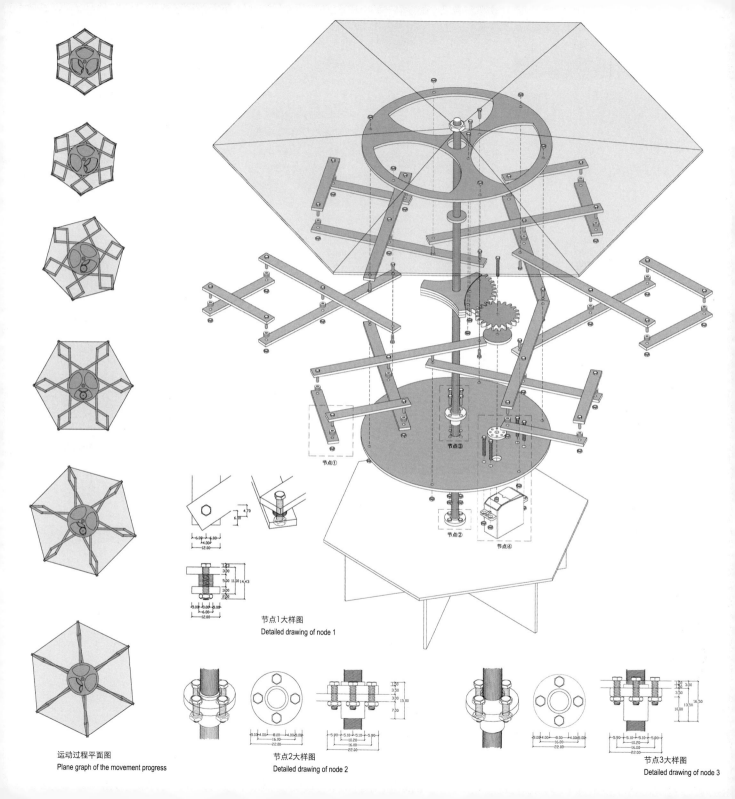

意大利旧建筑改造设计
RESTORATION OF HISTORIC ARCHITECTURE IN ITALY

毛里奇奥·德·维塔 华晓宁

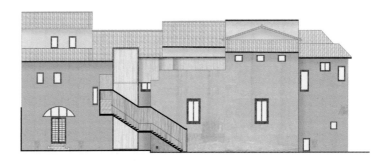

本次课题的关注点在于学习意大利建筑改造的方法,包括对旧建筑的历史考证、数字测绘,提出修复性意见及功能更新设计,要求对旧建筑不做过多干预。

建筑位于佛罗伦萨附近的一个小城市(Lastra a Signa),这座城市从建城开始就作为一个重要的交通枢纽,它位于阿诺河和山脚之间,如今佛罗伦萨到比萨之间的铁路也会途经这里。所需要改造的建筑位于整个小城中心,据历史考证,很有可能是建筑大师伯鲁乃列斯基最早年的作品,如今几近荒废。另外,这个城市还保留着部分老的城墙,这些老城墙由各个历史阶段的石材堆砌而成,有重要的研究意义和保存价值。基于这些原因,还考虑到当地居民对文化活动的需求,决定于该建筑赋予新的文化意义,即一座陈列当地古石的博物馆和当地居民可使用的图书馆。

目前,该建筑的底层部分得到修复,而上层尚未进行重大修复。该建筑曾用作教堂、医院、孤儿院、剧院、住宅等。如今则作为市政活动、临时展览、讲座或研讨会的场所,所有活动仅在底层举办,上层已不再投入使用。我们可以看到墙上有若干个历史悠久但未使用的壁炉或神龛,现有的楼梯也不足以满足通行需求,部分房间非常阴暗。主体左侧的三角区域难以使用,尽管正面非常有趣而且有助于人们了解关于石材的历史和材料。另外,由于部分历史原因,我们不允许改变建筑墙体和底层,初步的考虑是将该建筑改造成文化中心。

在研究过程中,我们发现这座城市有各种各样的石材。如果将其放在一起陈列将会非常有趣。这些石材可以来自城墙、城楼甚至是古建筑。这是该城市历史的重要组成部分。这里的石材不仅仅是石材,而是诉说着这座古城历史和故事。我认为这是在石材和历史之间建立一些联系的正确方式。

人们普遍认为历史枯燥乏味且难以理解。通过这种方式,我们可以很容易地接近城市的历史。除此之外,博物馆不仅仅意味着展览,还意味着可提供一些互动场所和活动。因此,我对修复伯鲁列斯基"原始"主体的最终决定是:两个房间用于展览,一个房间供游客对石材进行DIY。顺便提一下,现计划将另一个房间用作举办会议和讲座的场所。

This subject focuses on learning method for reforming old Italian architecture, including textual research of old architecture, digital surveying and mapping, proposing repair opinion and function updating design, old architecture shall not be intervened too much.

Lastra a Signa is a small city nearby Florence. As an important transportation hub since being established, it is located between Arno River and mountain foot, where the railway between Florence and Pisa runs through today. What we will reform is Spedale di S.Antonio, located in the center of the whole city, according to textual research, it is likely the early works of master architect Brunelleschi, almost abandoned today. In addition, some old city walls are still reserved in this city, built by stones at different historical stages, with important research meaning and reservation value. Basing on these causes, considering the requirement of local residents on cultural activity, it is decided to endow this architecture with new cultural meaning, that is, a museum that displays local old stones and library for local residents.

Currently the structure is restored in the part on the ground floor, while the upper floor has not undergone significant restoration. It was used for church, hospital, orphanage, theater, housing and so on. Nowadays it provides space for municipal events, temporary exhibitions, lectures or workshops. All activities are held only in the ground floor, the upper floor has gone out of use. We can see there are several historical but no-use fireplaces or tabernacle on the wall. The existing staircase is also not enough for traffic. Some rooms are very dark. The triangle area on the left of the body is hard to use, though the facade is interesting and helpful for people to read the history and materials of stones. What's more, we are not allowed to change the structure walls and the ground floor for some historical reasons. It's preliminarily considered to be converted to the cultural center in Lastra a Signa.

In the course of research, I found that there're many kinds of stones in this city. It will be interesting if we put them together to be on display. These stones can be from the city walls, the city towers or even the ancient buildings. It's a significant component of history of Lastra a Signa. Here stones are not just stones, but ones that are talking about the history and story of this old city. I think it can be a good way to make some connections between stones and history.

It's generally regarded that history is boring and hard to read. In this way, we can easily get close to history of Lastra a Signa. Beyond that, museums don't just mean exhibitions, but some interactive space and activities. So, two rooms for exhibitions and one for visitors to do some DIY stones is my final decision about restoration for the "original" body by Brunelleschi. By the way, the other room is planned to be the space for conferences and lectures.

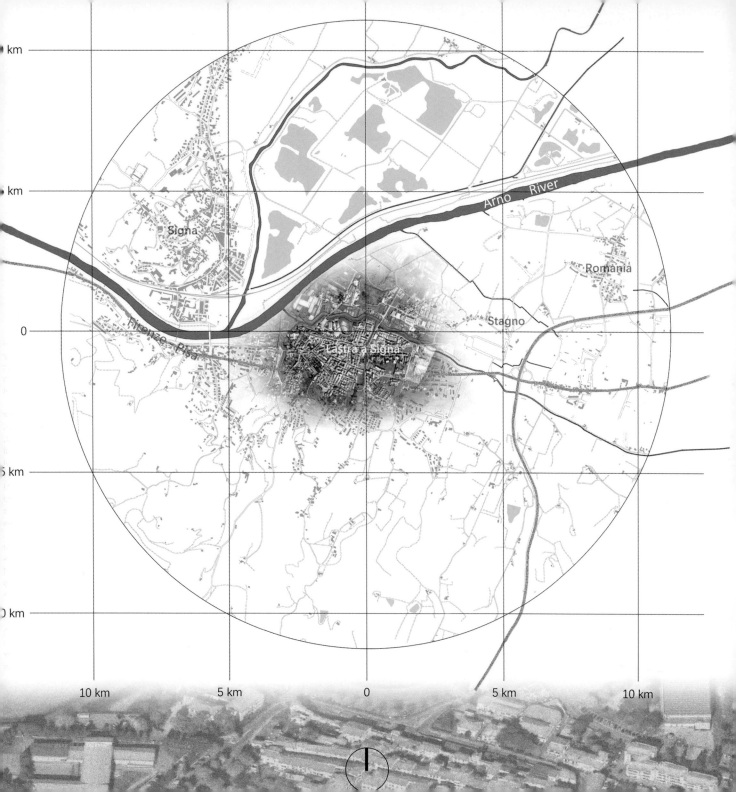

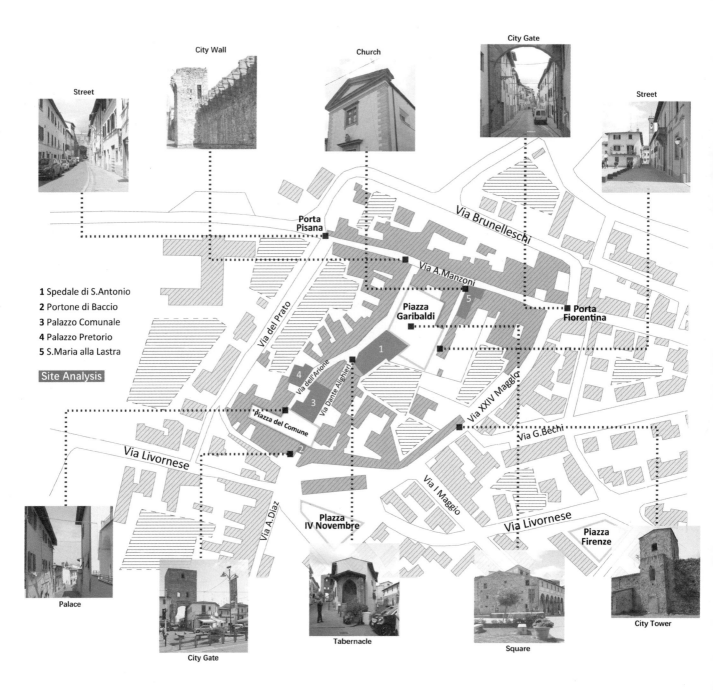

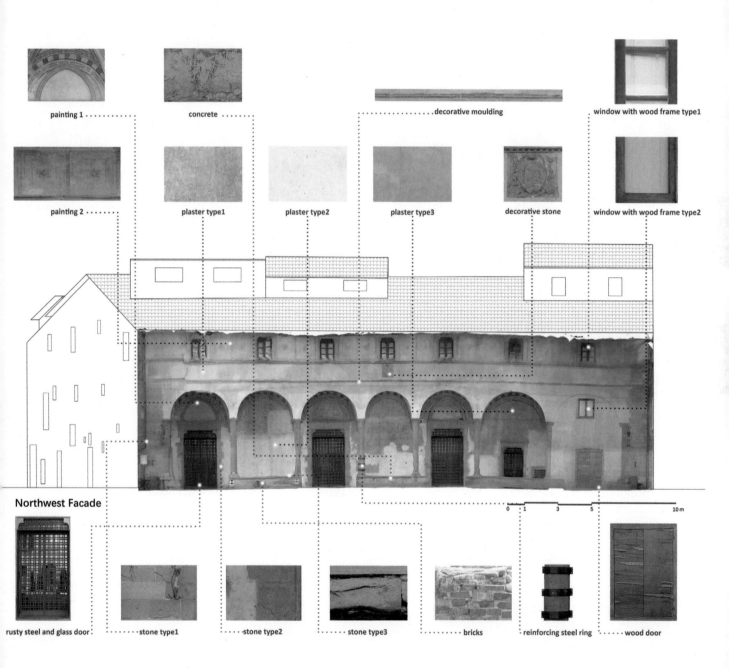

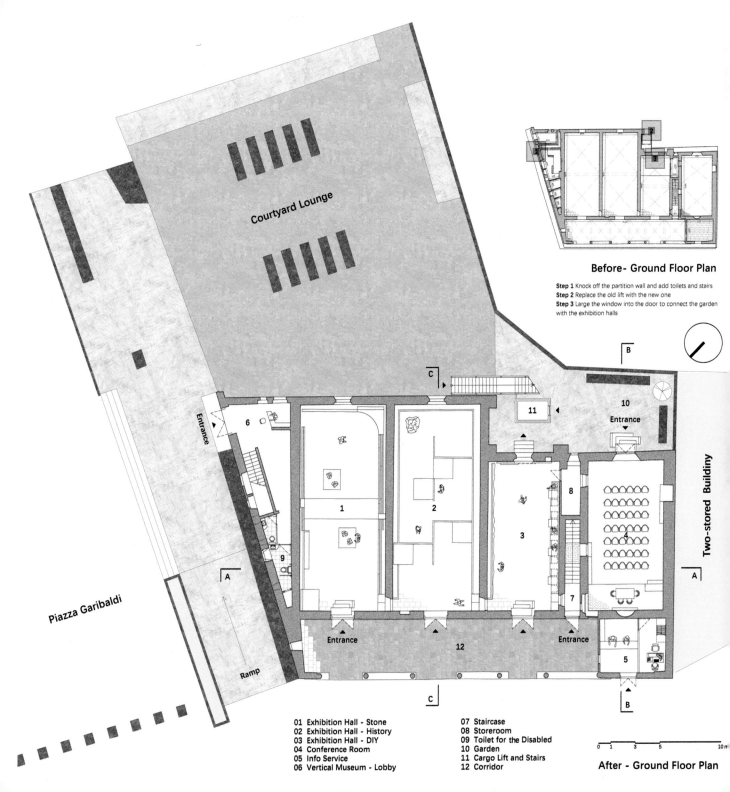

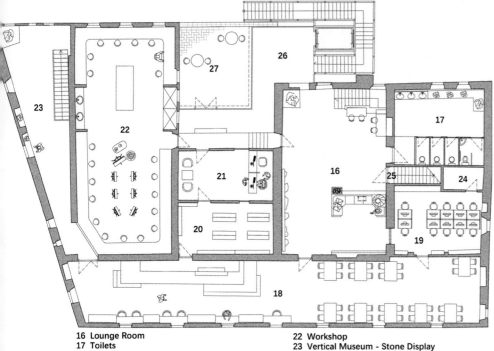

16 Lounge Room
17 Toilets
18 Library
19 Computer Room
20 Stack Room
21 Reception Office
22 Workshop
23 Vertical Museum - Stone Display
24 Storeroom
25 Staircase
26 Lift Hall with Rainshade
27 Terrace

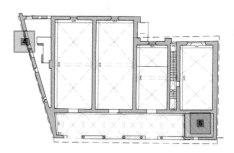

13 Vertical Museum - Wall Display
14 Double-high Space
15 Storeroom

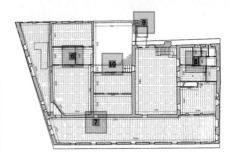

Before- First Floor Plan

Step 6 Knock off the partition wall for workshop
Step 7 Add stairs to connect the different levels
Step 8 Knock off the partition wall and redistribute the partition for toilets
Step 9 Make the lift and stairs reach the terrace
Step 10 Add rainshade on the terrace

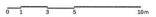

After - First Floor Plan

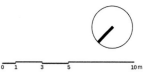

Before- Interlayer Plan

Step 4 Open partial floor to produce double-high space
Step 5 Make use of the interlayer room for storage

After - Interlayer Plan

65

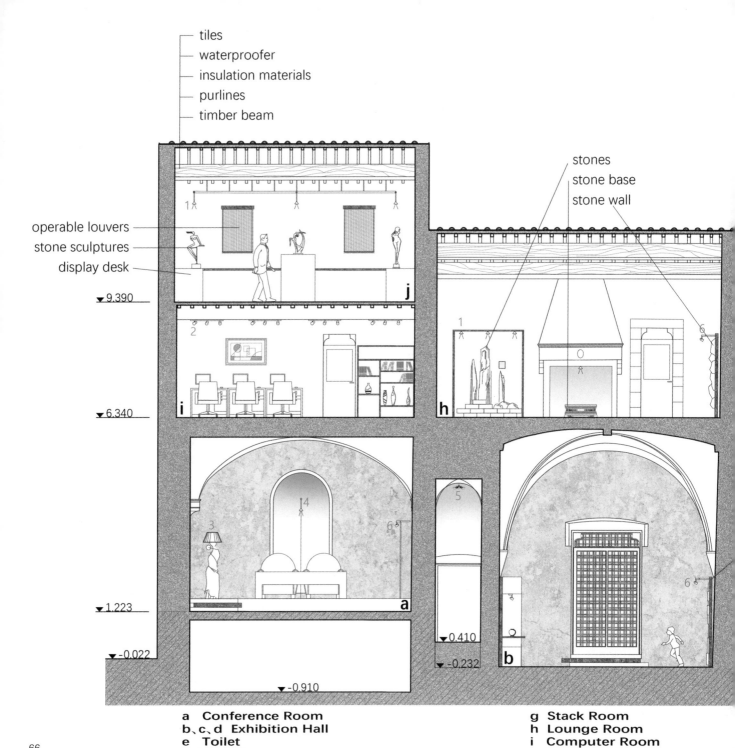

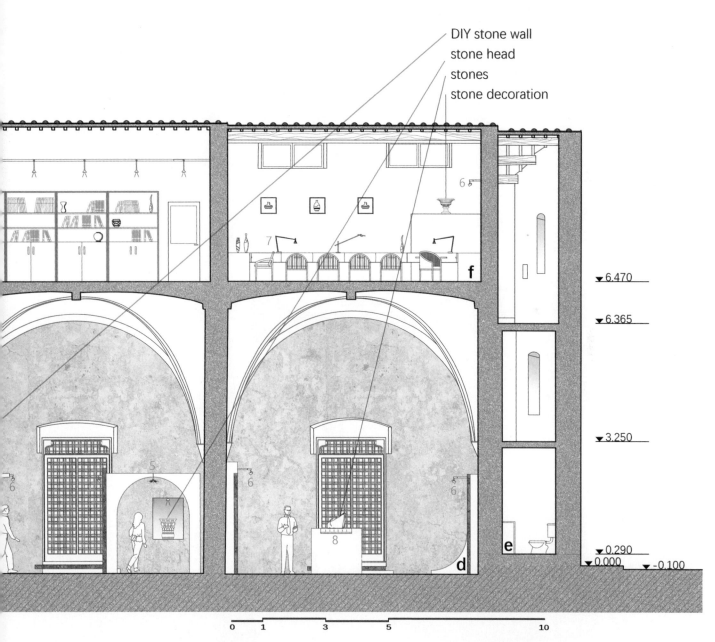

A - A Section

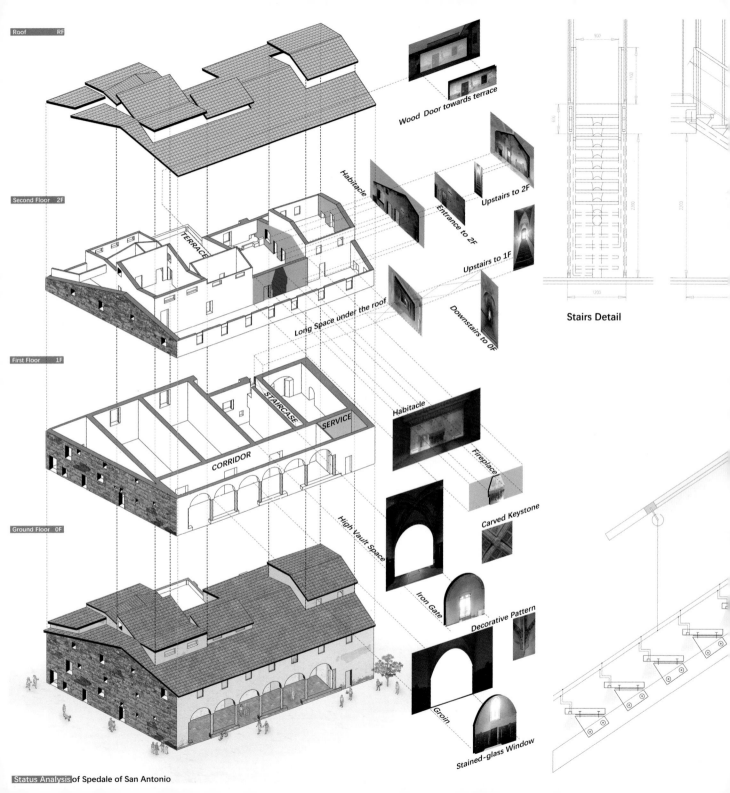

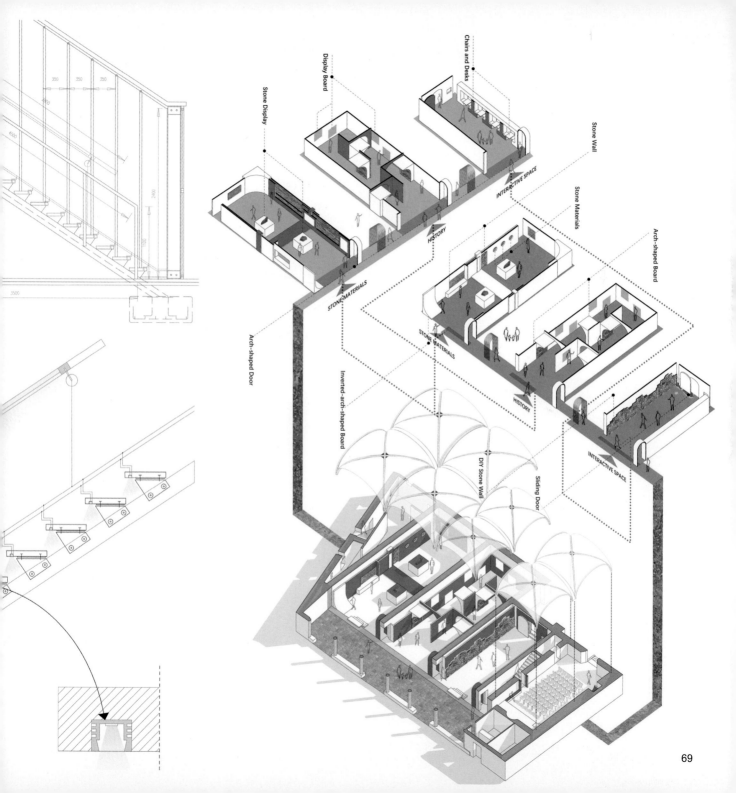

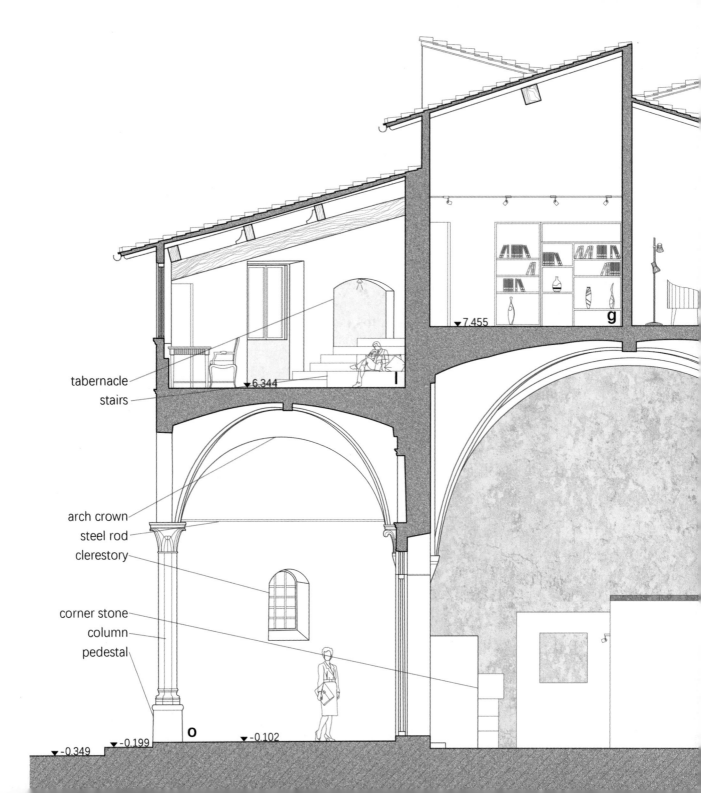

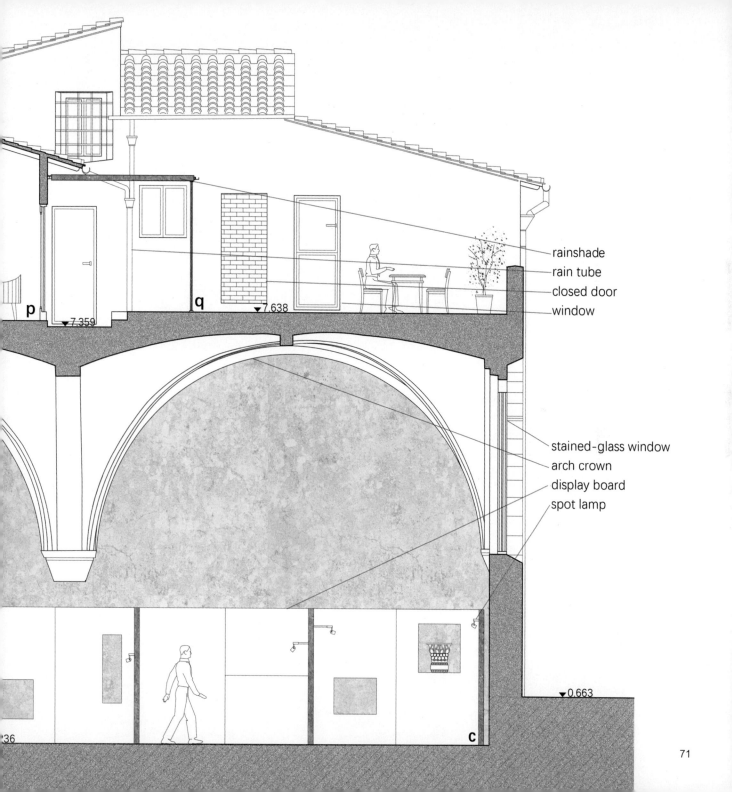

研究生国际教学交流计划课程
THE INTERNATIONAL POST-GRADUATE TEACHING PROGRAM

南京大学建筑与城市规划学院一直致力于最前沿的学术理论研究,并以之解决实际问题。在研究成果输出的基础之上,学院积极参与国际学术交流并致力为未来培养建筑和城市规划的高水平人才。为了拓展学院的国际学术交流网络并且建立长期的合作交流体系,在2016年南京大学建筑与城市规划学院已经启动了"研究生国际教学交流计划课程"(IPTP)。IPTP是一个灵活的邀请教学计划,每年都会有10位专家教授接受邀请前来教学访问。

今年开始,"研究生国际教学交流计划课程"正式实施。已经完成的课程有:
(1) 可持续环境设计:适用于跨文化案例研究
　　胡塞·阿尔莫多瓦、唐莲
(2) 自适应接头:运用3D打印接头建造居住建筑
　　朴大权、钟华颖
(3) 结构交互式设计在可持续发展设计中的应用
　　凯瑟琳·德·沃尔夫、孟宪川
(4) 数字技术:可快速搭建的木构亭设计
　　马库斯·哈德特、孟宪川
(5) 空间摩擦:城市设计思想作为设计策略01
　　米格尔·真蒂尔、傅筱
(6) 空间摩擦:城市设计思想作为设计策略02
　　夏洛特·马特尔、傅筱
(7) 建筑技术设计工作营:预制生活单元
　　罗伯特·博洛尼亚、华晓宁
(8) 重新定位城市空间:南京社会空间潜力的映射
　　弗洛里安·科萨克、窦平平
(9) 可持续建筑设计:智慧建筑建造和能效研究
　　塞尔吉奥·G. 梅尔加、周凌

Since its foundation, the School of Architecture and Urban Planning, Nanjing University is committed to cutting-edge researches and original theoretical contributions, in order to address contemporary issues. Based upon research outputs, the school is actively engaged in international academic exchanges and targets at nurturing high-level professionals in architecture and urban planning for the future. The school is actively engaged in international academic exchanges. In order to further extend its international academic exchange network and to establish a long-term cooperative and exchange mechanism, the school has launched the International Post-graduate Teaching Program (IPTP). The IPTP is a flexible guest-teaching program that includes 10 visiting positions annually.
The IPTP has been carried out since this year. Programs finished include:
(1) Sustainable Environmental Design: Application to Cross-Cultural Case Studies
　　José M. ALMODÓVAR、TANG Lian
(2) Adaptive Joints: Constructing Habitable Structures Using 3D Printed Joinery
　　PARK Daekwon、ZHONG Huaying
(3) Interactive Structural Design: Applied to Environmental Buildings
　　Catherine de WOLF、MENG Xianchuan
(4) Digital Technology: Easy-to-Assemble Interlocking Timber Pavilion Design
　　Markus HUDERTR、MENG Xianchuan
(5) Spatial Frictions: Urban Design Strategies as Tools 01
　　Miguel GENTIL、FU Xiao
(6) Spatial Frictions: Urban Design Strategies as Tools 02
　　Charlotte MALTERRE、FU Xiao
(7) Architectural Technology Design Studio: Prefabricated Living Unit
　　Roberto BOLOGNA、HUA Xiaoning
(8) Re-appropriating the Urban Realm-Mapping: Socio-Spatial Potentials in Nanjing
　　Florian KOSSAK、DOU Pingping
(9) Sustainable Architectural Design: Smart Building Construction and Energy Efficiency Research
　　Sergio G. MELGAR、ZHOU Ling

研究生国际教学交流计划课程 IPTP

可持续环境设计：适用于跨文化案例研究
SUSTAINABLE ENVIRONMENTAL DESIGN: APPLICATION TO CROSS-CULTURAL CASE STUDIES

胡塞·阿尔莫多瓦　唐莲

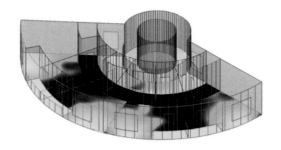

阿尔莫多瓦教授指导的"环境交互设计"工作营向学生介绍了如何设计可持续的建筑，以及如何在设计过程中使用不同的环境设计工具。通过对物理环境/场地、形式、体块/朝向、室内构成、外部围合的效果的整体分析和性能模拟，课程提供了跨学科及跨文化的环境来学习和探索新的设计概念以及策略。环境设计的概念和方法将引领整个设计过程。在为期两周的工作和研讨中，学生们在阿尔莫多瓦教授的指导下探索如何分析与评估并改善建筑的环境性能，最终取得较好的设计成果。

课程包括两个部分：讲座环节和工作坊环节，强调从做中学。在讲座环节，阿尔莫多瓦教授引导学生充分理解传统建筑材料、形式与空间组织与环境效能的关系之后，又向同学们介绍了Home Energy Efficient Design（HEED）、CLIMATE CONSULTANT、SEFAIRA、ECOTECT等软件，指导学生如何在设计过程中对建筑的天然采光及围护结构传热性能等进行评估与分析。其中HEED和CLIMATE CONSULTANT是加利福尼亚大学洛杉矶分校（UCLA）开发出来的建筑能耗和气候条件分析工具。

在工作坊环节，每个学生被要求选择一个建筑案例作为设计优化对象，先利用CLIMATE CONSULTANT分析建筑所在区域的气候条件，确定该建筑的优化目标，然后通过利用SEFAIRA、ECOTECT软件来分析建筑案例中存在的天然采光、热工性能问题，最后提出解决方案。阿尔莫多瓦教授在研究的基础上总结了建筑围护结构各立面的设计要点，并教授同学们用图解的方式去思考优化采光及日照问题。

The "Environmental Interactive Design" workshop, directed by professor José Almodóvar introduces students how to design sustainable buildings and to use different environmental design tools. It provides a cross disciplinary environment and cross cultural environment for learning and exploring new design concepts and strategies through integrated analysis and performance simulations of effects of physical environment and site, form, massing and orientation, internal configuration, external enclosure. The concept and methods of the environmental design will guide the whole design process. In the two-week work and study, under the direction of professor Almodóvar, the students explored how to analyze and improve the building environmental performance, and finally achieved better design works.

The course includes two parts: lecture series and workshop that emphasis on learning by doing. After guiding students to fully understand the relationship between traditional architecture's material and configuration and environmental efficiency, professor Almodóvar introduced environmental design tool, including Home Energy Efficient Design (HEED), CLIMATE CONSULTANT, SEFAIRA, ECOTECT and directed the students to carry out building's daylighting and thermal performance in the design process, with the help of the analysis tool mentioned above. Among these tools, HEED and CLIMATE CONSULTANT are developed by University of California Los Angeles (UCLA), which can respectively assess building energy and weather condition.

In the following teaching process, each student was required to choose a construction case as the object to optimize. Firstly, using CLIMATE CONSULTANT to analyze the climatic conditions where the building located, and then, the optimal goal of the building should be determined, after that, analyzing the existing problems—natural lighting and the thermal performance of this building—by using SEFAIRA and ECOTECT. Finally, the solution was put forward according to the analysis above. On the basis of the research, professor Almodóvar summarized the main points when design each facade of the building envelope, and also taught the students to think how to optimize the daylighting and sunshine problems in a graphic way.

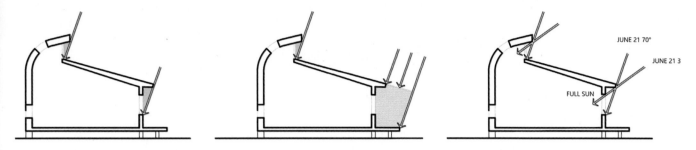

Window overhangs(designed for this latitude) or operable sunshades (by using deciduous plants) can reduce or eliminate air conditioning

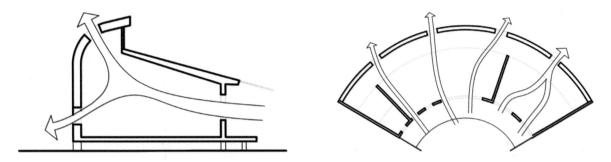

Use open plan interiors to promote natural cross ventilation, or use louvered doors, or instead use jump ducts if privacy is required.

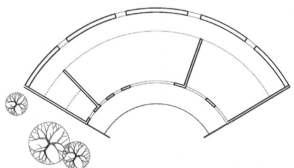

Use plant ,materials (bushes, tress, ivy-covered walls) especially on the west to minimize heat gain (if summer rains support native plant growth)

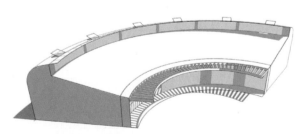

Minimize or eliminate west facing glazing to reduce summer and fall afternoon heat gain.

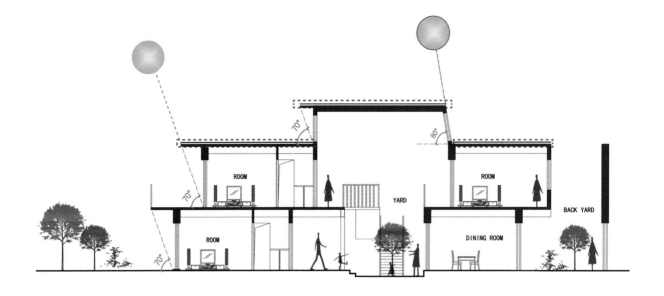

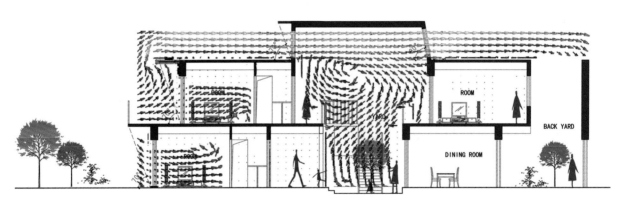

1-1 Section Flow Factor (Summer)

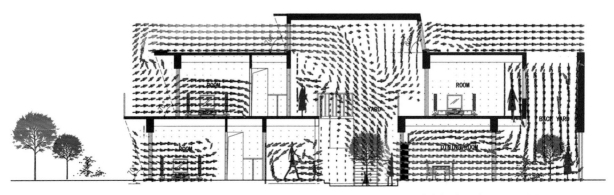

2-2 Section Flow Factor (Summer)

研究生国际教学交流计划课程 IPTP

自适应接头：运用3D打印接头建造居住建筑
ADAPTIVE JOINTS: CONSTRUCTING HABITABLE STRUCTURES USING 3D PRINTED JOINERY

朴大权 钟华颖

"自适应接头"讲习班用全新的方式探讨可连接传统材料（织物、木材/塑料/纸板、木棍、网格和管道等）的3D打印接头的应用前景。学生将学习使用这种接头设计和建造可居住的结构（例如展馆、遮阳篷、隔墙或家具等）。课程内容包括学习柔性、不稳定性、适应性和编程等，还将探讨柔性、双稳态、剪纸艺术、折纸、软/柔性接头、混合接头等机理。

学生通过在讲习班的学习，不仅能够学到数字设计的技能并了解物品的制造过程，而且还能学习如何将高等几何和构造学应用到设计材料系统上。讲习班将引入新兴的数字技术和加工过程，包括衍生设计、几何优化和编辑脚本等，并要求学生动手操作，学生将利用学到的知识完成其设计研究项目。

课程包括两个部分：新节点探索与新结构设计，强调从多领域、多学科思考建筑节点的可能性。在讲座环节，朴大权教授向学生介绍了一种"自下而上"的建筑设计方式——关注和研究一种新型材料或新机械结构的性能和特性，探寻出一些具有普适性和应用可能性的组合方式，并择选其中一部分进行深化，从而更好地完成建筑设计。课程向学生展示了诸如从昆虫翅膀的微观结构、材料不同泊松比的特性、柔性材料不同充气状态的形态变化等科学现象和原理所引发的新建筑节点设计。朴大权教授还引导学生从互锁结构、兼容性机械、双稳态结构、折纸结构、剪纸结构、屈曲特性、柔性材料、复合材料、可编程结构这几种方向出发，进行探索设计。

The "Adaptive Joints" workshop explores the potentials of 3d printed joints that can connect conventional materials (fabric, wood/plastic/ paper boards, sticks, meshes, and pipes) in novel ways. Using this joint design, the students will design and construct a habitable structure (e.g., pavilion, canopy, partition, or furniture). Concepts including flexibility, instability, adaptability, and programmability will be introduced; and mechanisms such as compliant mechanisms, bi-stable mechanisms, kirigami, origami, soft/flexible joints, and hybrid joints will be investigated.

Through the workshop, the students will not only be able to develop skills and knowledge of the digital design and fabrication process, but also learn how to apply advanced geometry and tectonics in design material systems. Emerging digital techniques and processes including generative design, geometry optimization, and design scripting will be introduced as hands-on workshops, and the students will utilize them to develop their design research project.

Our courses consists of two parts: new node exploration and new structural design,both emphasizing the possibility of building nodes in multiple fields and multiple disciplines. During the lecture session, Prof. Park Daekwon introduced students to a "bottom-up"approach in architectural design—focusing on and studying the properties and characteristics of a new type of material or new mechanical structure to explore some universal applications which are possible in utilization. We then selected the best application for better completing the architectural design. The course presented students with new construction node designs such as the microscopic structure of insect wings, the different Poisson's ratio of materials, and the morphological changes of different inflatable states of flexible materials. Prof. Park Daekwon also provided students with guidance of different directions for better exploring the design.Typical of them are interlocking structure, compatible machinery, bi-stable structure, origami structure, paper-cut structure, buckling properties,flexible materials, composite materials, programmable structure.

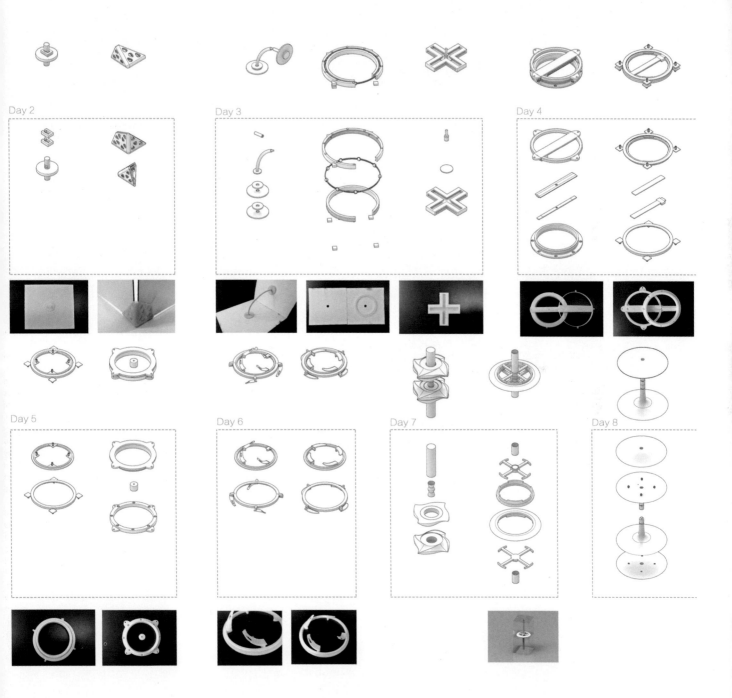

79

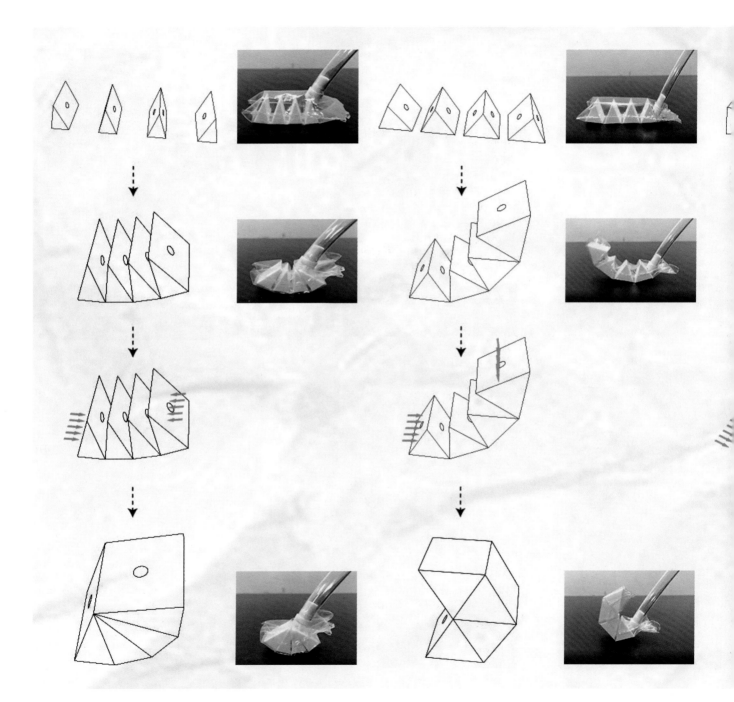

研究生国际教学交流计划课程 IPTP

结构交互式设计在可持续发展设计中的应用
INTERACTIVE STRUCTURAL DESIGN APPLIED TO ENVIRONMENTAL BUILDINGS

凯瑟琳·德·沃尔夫 孟宪川

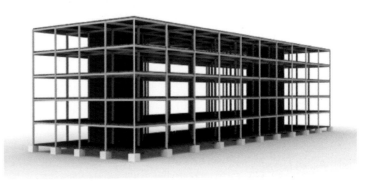

本课程向学生介绍互动式结构设计,重点介绍循环经济原理。建筑的循环经济原理是指通过设计将目前的采掘—生产—处置这一工业模式转变为可恢复和可再生的建筑环保模式。我们只有在结构设计实践中增加第四个维度:时间维度,才能减少建筑物对环境的影响。四维(4D)结构设计意味着全系统的创新,通过积极的社会利益重新定义增长。因此,本课程将从材料提取、生产、运输到工地、施工、维护和建筑结构拆除等方面介绍降低温室气体排放的全行业策略。交互式结构设计通过构建信息模型,可实现创建环保建筑的目标。

教员将把学生分成不同的工作团队,并给每个团队分配不同的研究案例。教员还会给每个团队分发中国建筑物数量清单和建筑信息模型。学生在经过数据分析和计算碳消耗后可以评估他们的项目对环境的影响程度,并可将结果与其他团队进行比较。学生需要做一个简单的跨度为6m的跨越结构设计练习,并共同讨论建筑业的行业策略和结构工程师在建筑结构走向循环经济过程中的驱动力作用。

This course introduces students to structural interactive design, with a focus on circular economy principles. Circular economy of buildings shifts the current extract-produce-dispose industrial model towards a restorative and regenerative built environment by design. Reducing the environmental impacts of buildings can only occur if we add a fourth dimension to structural design practice: the time dimension. Four-dimensional (4D) structural design implies system-wide innovations that redefine growth through positive societal benefits. The class therefore introduces industry-wide strategies to lower greenhouse gas emissions related to material extraction, production, transport to the site, construction, maintenance and demolition of the structure of buildings. Interactive structural design through building information modelling will be illustrated to achieve the goals of environmental buildings.

Students will be distributed in working teams and different case studies will be assigned by the professor to each team. The bill of quantities and building information models of a building in China will be given to each team. Data analysis and embodied carbon calculation will then allow the students to assess the environmental impact of their project and compare it to other groups. Students will be given a design exercise for a simple structure to span 6 m. Meanwhile, the industry strategies and drivers for the construction industry and structural engineers to move towards a circular economy in building structures will be discussed with the students.

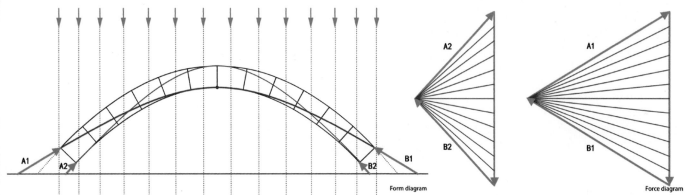

Form diagram · Force diagram

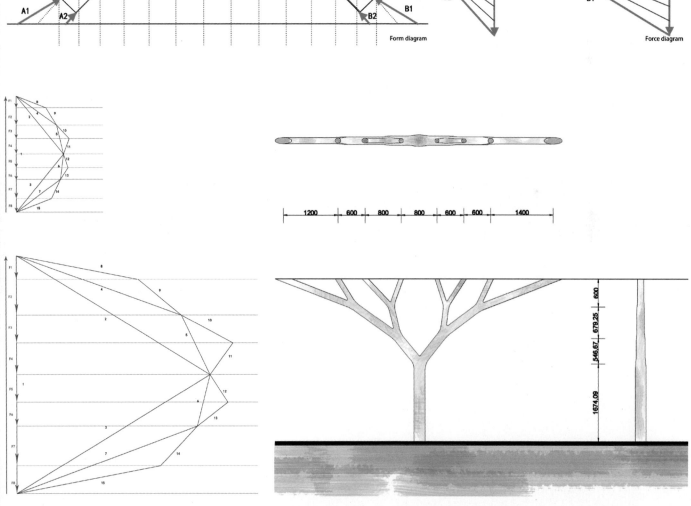

研究生国际教学交流计划课程 IPTP

数字技术：可快速搭建的木构亭设计
DIGITAL TECHNOLOGY: EASY-TO-ASSEMBLE INTERLOCKING TIMBER PAVILION DESIGN

马库斯·哈德特　孟宪川

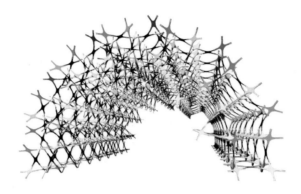

本课程探讨数码工具在非传统建筑结构和材料系统的设计和制造中的应用前景。该体系中的结构性能和空间质量特别有意思。学生将运用在本课程学到的知识，设计一个临时的展馆，并在南京大学校园内进行制作和组装。本课程适合计算程序设计的初学者，它将传授参数设计、材料计算和数字制造等基础理论。

课程第一周介绍数字设计和加减制造技术的基础理论，会同时涉及理论和实践两个方面。学生们在每次半天的讲习班上，将学习素描、绘图以及实物和数字模型，了解基本的装配策略和构造系统以及它们的结构和空间质量。学生在学习这些模拟和数字课程后，将可单独或两人一组制订一个设计方案。课程第二周，学生将改进他们的设计方案，并准备其材料等待中期检查。检查将于星期四下午进行。我们将挑选出一个设计方案，作为整个组进一步推进的设计方案。一到两段简短的文字说明可以帮助学生陈述他们的设计方案，也便于准备报告。学生将在课程的最后阶段，一起合作制作和组装一个临时展馆。他们还要为了展馆的开馆准备海报和记录展馆设计和制作过程的小册子。

本课程采用自下而上的教学方法，这就意味着，整体建筑架构清晰度和用途不是先决条件，而根据材料系统的属性构建建筑才是关键。本课程的出发点是办一个每次半天的讲习班，学生可以在此运用简单的几何元素亲自动手，学习基本的装配策略。实物和数字建模在整个课程中起着举足轻重的作用。

This course explores the potential of digital tools in the design and fabrication of non-conventional architectural structures and material systems. The structural performance as well as the spatial qualities of such systems are of particular interest in this context. Based on these aspects, the participants will design a temporary pavilion that will be fabricated and assembled in the compus of Nanjing University. The course aims at beginners in computational design and introduces the basics of parametric design, material computation and digital fabrication.

The first week starts off with an introduction to digital design and both additive and subtractive fabrication techniques. Both theoretical and practical aspects are addressed. During a half-day workshop, and working with sketches, drawings and physical as well as digital models, the students will study basic assembly strategies and tectonic systems, as well as their structural and spatial qualities. Based on these analog and digital studies, they will start to develop a design proposal, either individually or in groups of two.During the second week, the students will refine their design proposals and prepare their material for an intermediate review. This review will take place on Thursday afternoon. One design proposal will be selected and further developed by the whole group. One or two short input lectures will support the students in developing their designs and in preparing for the presentations.In the final phase of the course, students will collaborate in the fabrication and assembly of a temporary pavilion. For the opening of the pavilion, the students will prepare posters and a booklet that documents the development process.

This course pursues a bottom-up approach, meaning that the overall architectural articulation and the purpose of use are not prerequisite, but result from the properties of the material system. The starting point of the course is a half-day workshop, in which the students study basic assembly strategies in a hands-on manner, using geometrically simple elements. Physical and digital modelling play a pivotal role throughout the course.

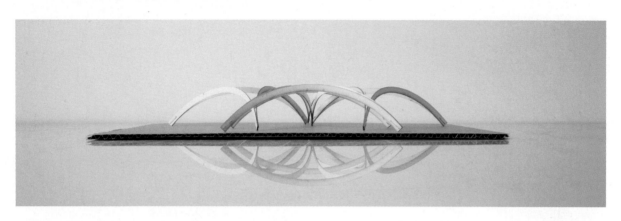
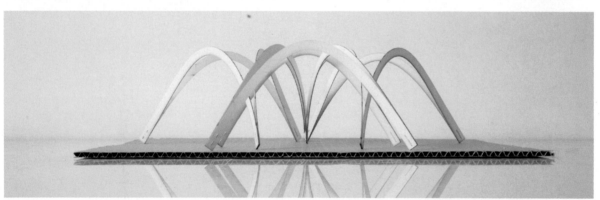
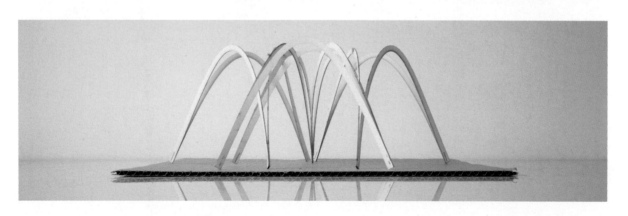

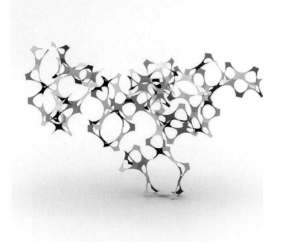
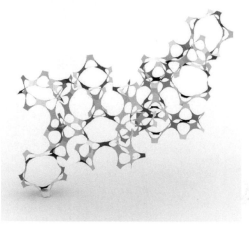
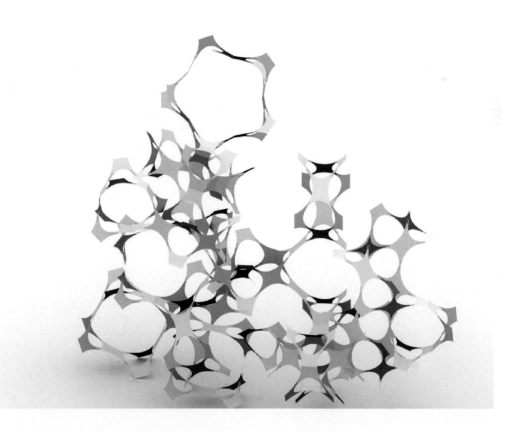

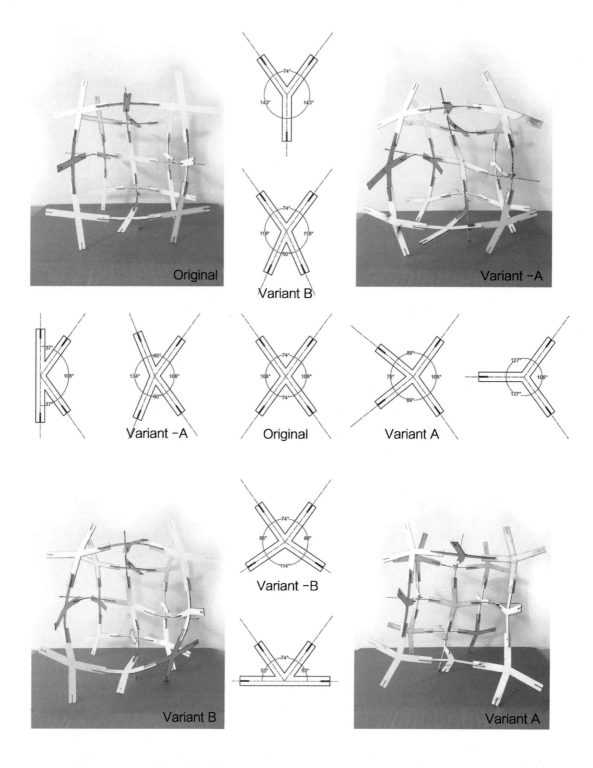

研究生国际教学交流计划课程 IPTP

空间摩擦：城市设计思想作为设计策略01
SPATIAL FRICTIONS: URBAN DESIGN STRATEGIES AS TOOLS 01

米格尔·真蒂尔 傅筱

中国城乡之间的互动非常具体。大规模的城市化已经越来越多地吸收农村地区，产生混杂的情况。城乡之间的边界确实是无处不在的，它属于中国新城市的DNA。工作营的目的是探索和提出一系列基于空间摩擦视角的方法，这些方法可以将建筑师和城市规划者重新定位为能够理解建筑物之外空间的行动者，并了解工作中的机制。

首先课程通过理论介绍、案例学习及实地调研，使学生们对"空间摩擦"这个概念有了初步的了解；接着通过一系列的调研及采访，学生们根据调研结果，自拟任务书，通过相关策略去解决调研中发现的问题。在为期一周的工作和研讨中，学生们在真蒂尔先生的指导下探索在城市化急剧发展的背景下，城市边缘区的农村环境将如何发展。

真蒂尔先生从"人"的角度出发，讲述城市与农村的关系。随着城市的发展，传统的农村将发生改变，农村人口的生活方式将发生改变，农村的传统功能也将发生改变。在这些过程中，"空间摩擦"同时发生，这种摩擦既有冲突与差异，也包含协同与合作。此外，真蒂尔先生向学生们介绍了几个经典的案例，来体现"空间摩擦"的特点。

通过本次工作营的学习，参与的同学们从不同的方面了解到当下中国城市化的进程，从人文方面及城市角度去思考农村的问题，提出相关的解决方案。城市与农村之间的"空间摩擦"关系可能并不是一种传统方式上"消灭"关系，尤其对成千上万的农村人口来讲，当他们变为城市人口时，他们传统的生活方式及居住环境可能会有一种更和谐的关系来呈现。

The interaction between urban and rural China is very specific.Large-scale urbanization has increasingly absorbed mixed conditions in rural areas.The border between urban and rural areas is indeed ubiquitous. It belongs to the DNA of China's new city.The purpose of the workshop is to explore and propose a series of methods based on spatial friction perspectives that can reorient architects and urban planners to actors who understand the space outside the building and understand the mechanisms at work.

The course begins with a theoretical introduction, case study and field research, so that the students firstly have a preliminary understanding of the concept of "space friction". Then through a series of research and interviews, the students self-designed the task program according to the survey results, and according to the related strategies to solve problems found in the survey.During the week-long work and seminars, under the guidance of Mr. Gentil, the students explored how the rural environment in the urban fringe will develop in the context of the rapid development of urbanization.

Mr. Gentil explains about the relationship between urban and rural areas from the perspective of "people". With the development of the city,the traditional rural areas will change, the lifestyle of the rural population will change, and the traditional functions of the countryside will change. At the same time as these processes are underway, "space friction" also occurs simultaneously. This friction has both conflicts and differences, as well as synergy and cooperation. In addition, Mr. Gentil also introduced several classic cases to the students to reflect the characteristics of "space friction".

Through the study of this workshop, the participating students learned from different aspects that they are facing the current urbanization process in China, thinking about rural issues from the perspective of humanities and cities, and proposing related solutions.The "space friction" relationship between urban and rural areas may not be a traditional way of "eliminating" relationships, especially for thousands of rural people, when they become urban populations, their traditional way of life and living environment may have a more harmonious relationship to present.

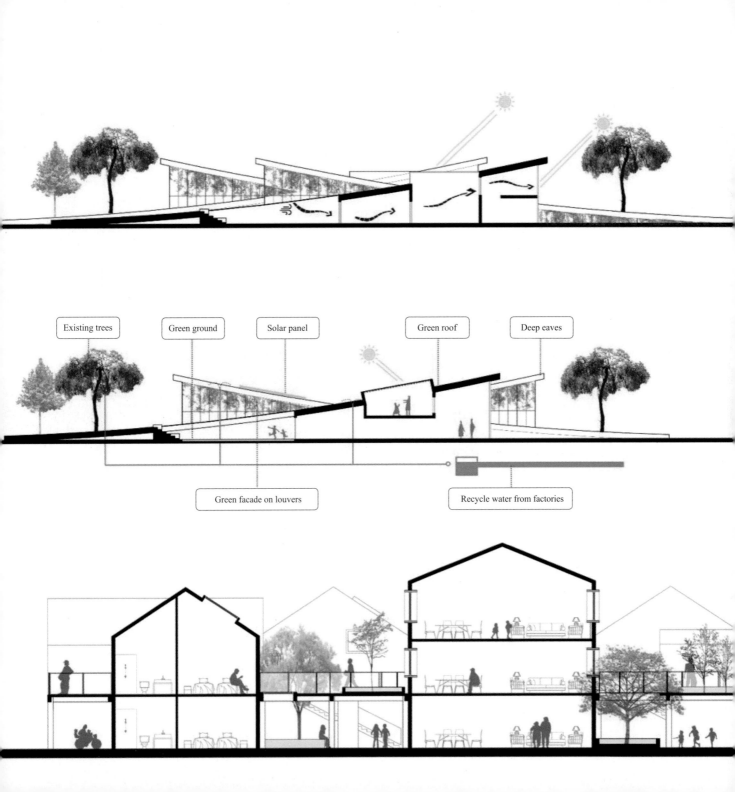

研究生国际教学交流计划课程 IPTP

空间摩擦：城市设计思想作为设计策略02
SPATIAL FRICTIONS: URBAN DESIGN STRATEGIES AS TOOLS 02

夏洛特·马特尔 傅筱

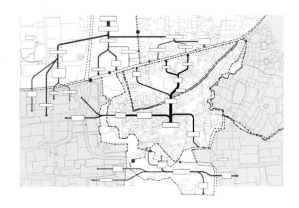

　　马特尔教授指导的"空间摩擦：城市设计思想作为设计策略"工作营于6月16日正式开启，并于6月23日圆满结束。"空间摩擦"课程试图探讨在中国城市化进程当中，城乡边界处发生的复杂的空间机制和摩擦关系。课程设置了"人口、资源、土地与管理、食物"4个主题，由4个小组各选取其一，并针对其中特定的问题进行研究。学生通过田野调查，寻找城市与乡村之间的空间摩擦关系，并形成一系列的图解，同时针对问题做出结论性的诊断，最终提出初步的城市策略层面的解决问题方案。

　　在本次课程中，学生不仅能够从文化、社会、经济、政治等多个角度对城乡边界关系进行深入理解并转化成概念性的图表，同时也能进一步地提出城市设计层面的发展策略。

　　课程包括3个部分：理论教学、场地调研和设计教学。首先，马特尔教授通过讲座向大家介绍了城市研究的参考和信息转化成图表的实例，并在之后一起进行实地调研。此后，学生将调研的发现与成果绘制成概念图表，提出针对特定问题的诊断并进行中期答辩。其次，夏洛特老师简要介绍了不同地区城市设计的发展策略。最后，学生根据各自的调研成果和中期反馈，提出概念性的城市设计策略。

The "Spatial Frictions: Urban Design Strategies as Tools" workshop, directed by Prof. Charlotte Malterre, began from 16th June, and successfully concluded on 23th June. The "spatial frictions" workshop tried to explicate the complex mechanisms and the frictions generated by the urbanization of China. "People, resources, land and governance, food" were set as 4 topics. Students would choose one and investigate through fieldwork and research urban and rural frictions. The goal is to represent their findings through mappings and diagrams, and out of these findings, to formulate a sort of atlas of spatial frictions. Based on these outcomes, an urban design strategy was to be developed.

Through this workshop, students could understand how these spatial frictions emerged in terms of cultural, social, economic and political mechanisms and translate them graphically. Furthermore, students would also draft an urban design strategy and a proposal for spatial intervention.

The workshop includes 3 parts: theoretical lecture, fieldwork and design process. At first, Prof. Malterre introduced references of urban research and how to represent findings in the form of "atlas", then all went to do fieldwork together. Based on the fieldwork, students drew maps, diagrams and axonometrical drawings of the spatial frictions they found, and proposed diagnoses for the mid-term presentation. Next step, Prof. Charlotte gave students several urban design strategy examples practiced all over the world. At last, students proposed the conceptual urban design strategies based on the research results and the feedback of the mid-term review.

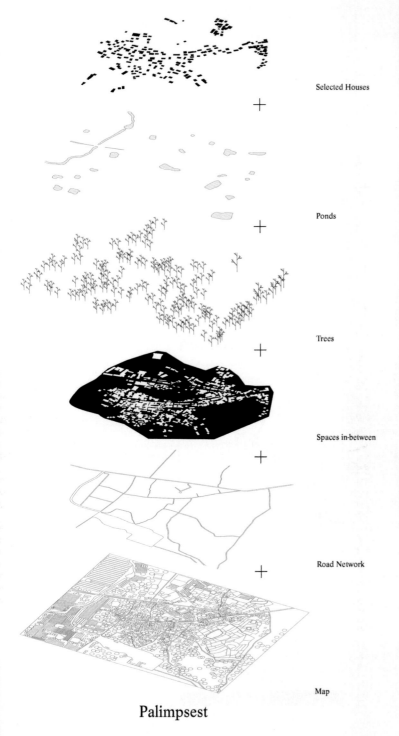

Selected Houses

Ponds

Trees

Spaces in-between

Road Network

Map

Palimpsest

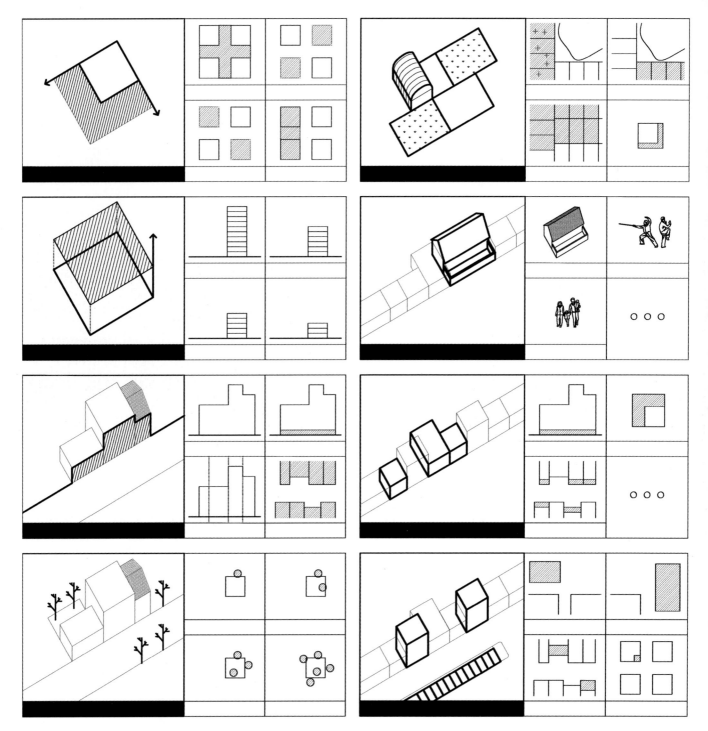

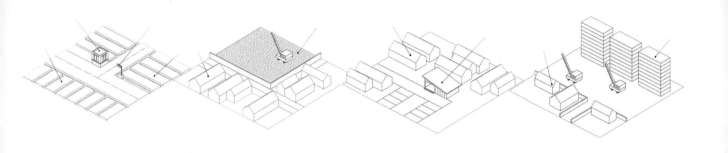

研究生国际教学交流计划课程 IPTP

建筑技术设计工作营：预制生活单元
ARCHITECTURAL TECHNOLOGY DESIGN STUDIO: PREFABRICATED LIVING UNIT

罗伯托·博洛尼亚 华晓宁

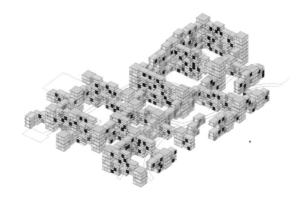

本课程主要向学生介绍如何根据住房和空间建筑原理，通过深刻地影响建筑的形式和概念的技术知识设计建筑单元。这种建筑单元设计着重于项目的"构造"维度，强调建筑方法和建筑体系，并把材料特性作为整个建筑表现形式的重要组成部分。建筑项目理念认为，构造是整个设计过程的一部分，因为各不同技术学科之间可平等互补。

本课程为学生提供了建筑设计的方法论和操作工具，着重于空间和技术系统以及与环境的关系。建筑技术的方法和工具将从可持续性、预制性、项目可行性、施工以及探索创新等原则出发，指导建筑的设计过程。

课程的主题是在一个城市环境中设计一个预制的最小生活单元。主题将根据"2018年UIA HYP杯国际建筑设计学生竞赛"设置的要求制定，作为该项目的背景。参加本课程的学生将根据教育目的完成建筑设计，同时还有机会在后期进一步完善并提交其项目设计用于参加比赛，以便实践他们的教育经验和评估比赛的方法。

本课程鼓励学生探究"微观建筑与宏观设计"之间的关系，这种关系也说明了空间审美与空间功能之间以及建造过程与产品设计之间的密切关系。

课程内容包括：
(1) 项目要求和功能空间规划；
(2) 预制方法及其含义；
(3) 最小生活单元的建筑类型学含义；
(4) 技术资源。

The course introduces students how to design a building unit according to the principle of architecture as inhabited and built space through a technical knowledge that profoundly affects the form of the project and its conception. The design studio focuses on the "tectonic" dimension of the project emphasizing the building methods and system, and takes the material character as significant part of the whole architectural expression, according to the concept of the architectural project as a whole design process based on the equivalence of all disciplines and complementarity between the different skills.

The course provides students with methodological and operational tools of architectural design, focusing on both the spatial and technological systems and the relationship with environment and context. The methods and tools of architectural technology will guide the design process considering the principles of sustainability, prefabrication, project feasibility and construction as well as exploring innovation.

The theme of the course is the design of a prefabricated minimal living unit in an urban context. The theme will be developed according to the requirements defined by the "UIA HYP Cup 2018 International Student Competition in Architectural Design" as a background of the project. Students attending the course will design their proposal for educational purposes but will also have the opportunity to complete it at a later time and submit their project for the competition in order to put in practice their educational experience and evaluate the contest's method.

The course encourages students to explore the dimension in between "micro-architecture and macro-design" which implies a close relationship among the aesthetic and function of the space, the construction process and the design of product.

Teaching contents deal with:
(1) project requirements and functional-spatial programme;
(2) prefabrication methods and implications;
(3) building typology of a minimal living unit;
(4) technological resource.

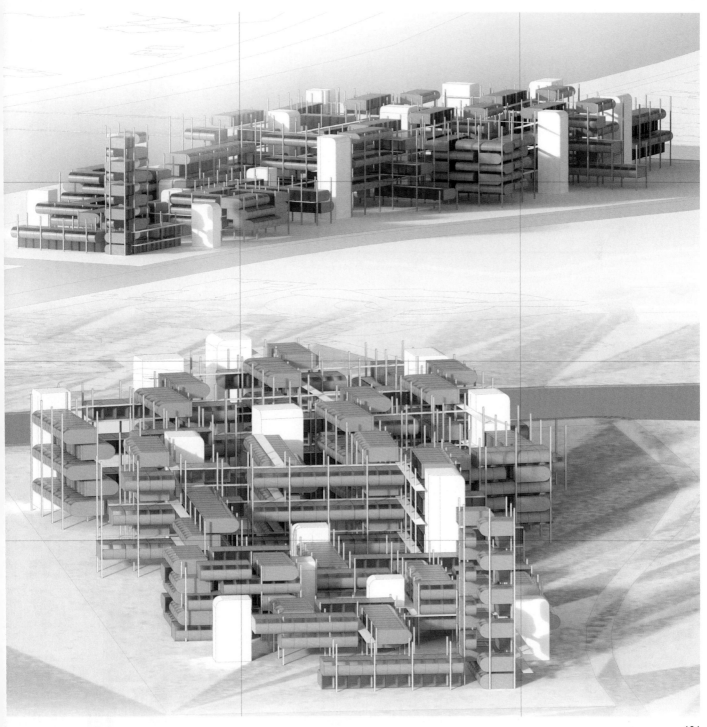

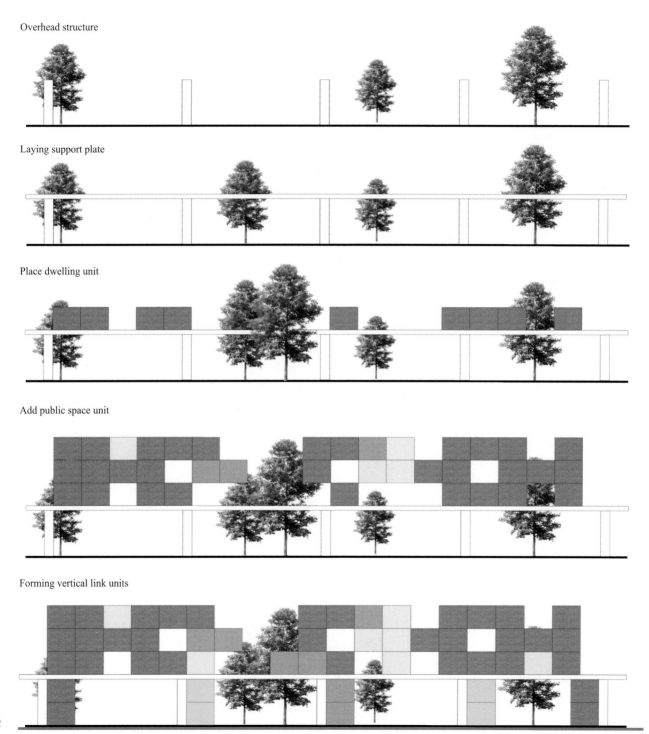

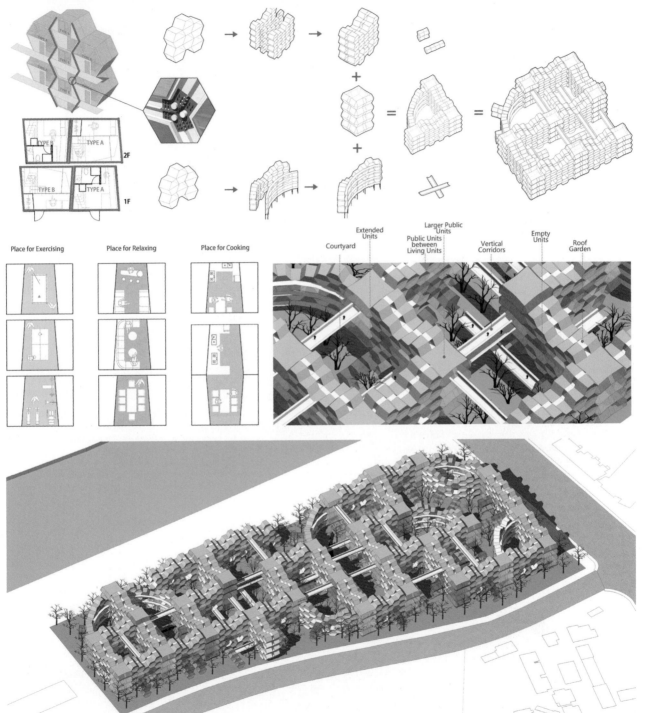

研究生国际教学交流计划课程 IPTP

重新定位城市空间：南京社会空间潜力的映射
RE-APPROPRIATING THE URBAN REALM-MAPPING: SOCIO-SPATIAL POTENTIALS IN NANJING

弗洛里安·科萨克 窦平平

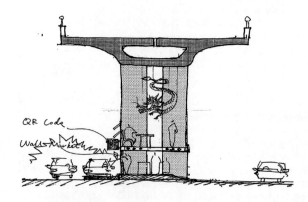

本次设计研究讲习班将探讨南京市日常生活情境的社会空间潜力。它将批判性地调查南京居民的居住设计策略，并探讨这些策略是如何成为可持续和公正的城市主义模型和催化剂的。我们将首先运用一系列观察和陈述的方法，根据在南京市内选定的城市状况绘制重新定位的社会空间图，这些方法会帮助我们客观地反映情况，还有助于我们提出自己的主张。本课程的目的是制订一套思考方案和提出主张的方法，探索如何通过干预策略和新空间实践来重新调整城市(甚至是"繁荣城市")的空间。

南京大学的学生将与谢菲尔德大学城市设计专业的硕士研究生一起参加这个课程的学习。这个讲习班是南京大学、CEPT大学艾哈迈达巴德、约翰内斯堡大学和谢菲尔德大学共同举办的国际教学和研究交流项目RAUM——研究替代城市方法的一个部分。

This design-research workshop will explore the socio-spatial potentials of everyday situations in Nanjing. It will critically investigate design tactics developed by the inhabitants of Nanjing and ask how these tactics can be seen as models and catalysts for a sustainable and just urbanism. We will start with mapping the socio-spatial of re-appropriation in selected urban situations within Nanjing, using a range of observational and representational methods that will help us to be both reflective and propositional. The aim is to produce a set of speculative and propositional approaches how to re-appropriate the urban realm (even in a "booming city") through tactical interventions and new spatial practices.

Students of Nanjing University will be joint by MA in Urban Design students from the University of Sheffield. This workshop is linked to the international teaching and research exchange project RAUM—researching alternate urban methods between Nanjing University, CEPT University Ahmedabad, University of Johannesburg and the University of Sheffield.

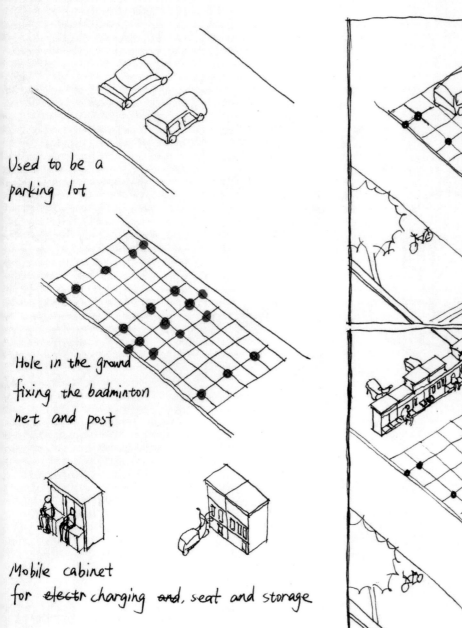

Used to be a parking lot

Hole in the ground fixing the badminton net and post

Mobile cabinet for ~~electr~~ charging ~~and~~, seat and storage

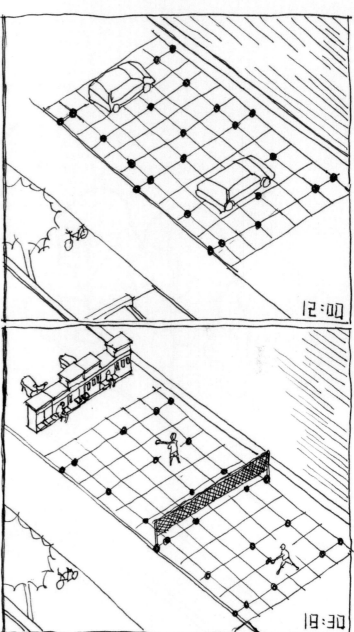

12:00

18:30

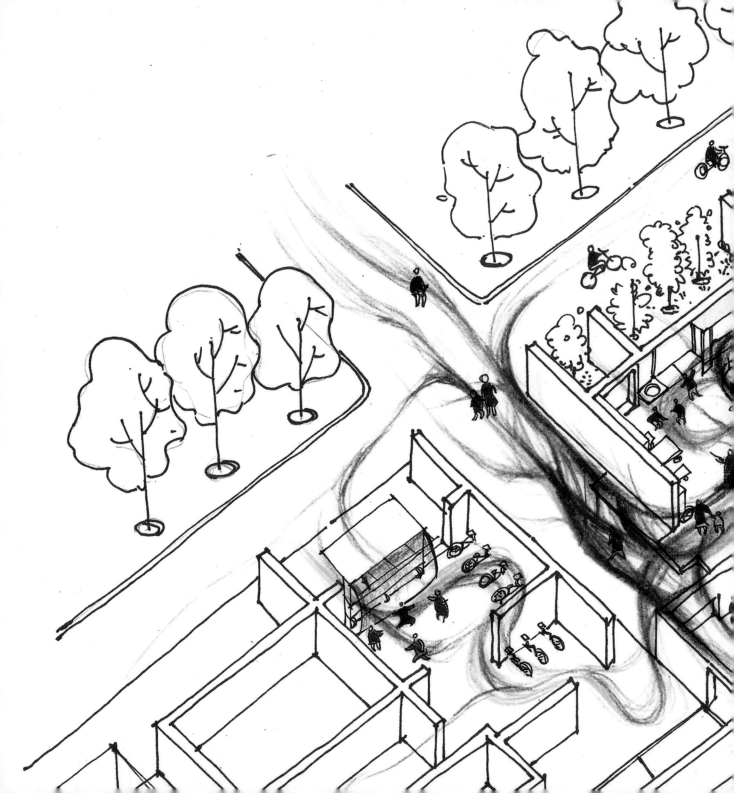

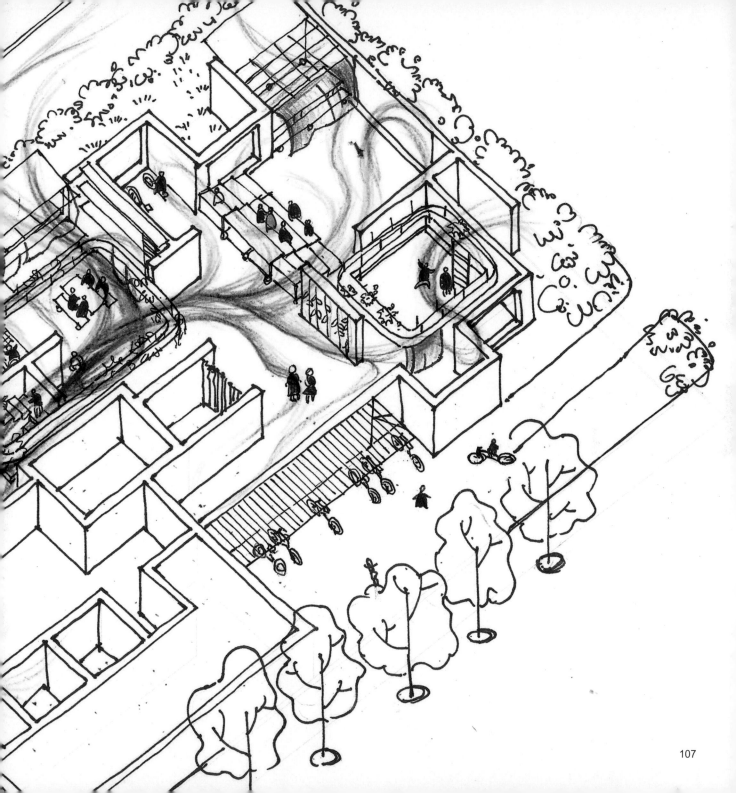

研究生国际教学交流计划课程 IPTP

可持续建筑设计：智慧建筑建造和能效研究
SUSTAINABLE ARCHITECTURAL DESIGN: SMART BUILDING CONSTRUCTION AND ENERGY EFFICIENCY RESEARCH

塞尔吉奥·G. 梅尔加　周凌

建筑能耗目前占据欧盟整体能耗的40%。由建造材料供给、建筑和城市全生命周期能耗所产生的温室气体（GHG），令建筑师成为应对全球变暖的重要行动者。建筑师群体不应再逃避责任。这意味着我们必须超越传统的舒适区：设计和形式。

我们生活于一个建筑实践范式转型的时代。如果说之前的现代主义运动关注社会住宅以及建筑与功能的关系，那么这一代和未来的使命则是融合建筑与环境，由此改善人类生活质量，应对全球性的挑战——当下，这个挑战就是气候变化带给人类的威胁。这种全新的建筑实践形式，需缝合建筑与工程的分离轨道以及两者与科学研究的关系。当代建筑设计不能仅仅理解为功能与形式的对话，而应整合工程学的解决方案，致力于将各自的成果传递给其他建筑师，完成螺旋上升的整体进步。

最低能耗建筑的概念不仅应用于新建筑，在现有建筑的改造及能耗修复工作中更为重要。原因是那些当代大型园区——大部分都是在节能规定颁布之前建造的——正是建筑领域最大的耗能单元，因此也拥有最大的减排潜力。即便不论节能问题，这些建筑往往也是不舒适的。创造性的外围护改造设计策略不仅适用于能量修复领域，也需探索新功能的整合、可再生能源的建筑综合表达。

Building energy consumption currently accounts for 40% of the EU's overall energy consumption. Greenhouse gases (GHG) generated by building materials supply, construction and urban life cycle energy consumption make architects an important actor in response to global warming. The group of architects should no longer evade responsibility. This means we must go beyond the traditional comfort zone: design and form.

We live in an era of architectural practice paradigm shift. If the previous modernist movement focused on social housing and the relationship between architecture and function, then the mission of this generation and the future is to integrate architecture and the environment, thereby improving the quality of human life and coping with global challenges—now, the challenge is climate change poses a threat to humanity. This new form of architectural practice requires the intergration of the separation of architecture and engineering, and the relationship between the two and scientific research. Contemporary architectural design can not only be understood as a dialogue between function and form, but should integrate engineering solutions and commit to transfer their results to other architects to complete the overall progress of spiraling.

The concept of the lowest energy building is not only applied to new buildings, but also more important in the renovation of existing buildings and energy restoration. The reason is that large contemporary parks—mostly built before the energy regulations were enacted—are the largest energy-consuming units in the construction sector and therefore have the greatest potential for emission reductions. Even regardless of energy-saving issues, these buildings are often uncomfortable. The creative peripheral regeneration design strategy is not only suitable for the field of energy restoration, but also for the integration of new functions and the comprehensive expression of renewable energy.

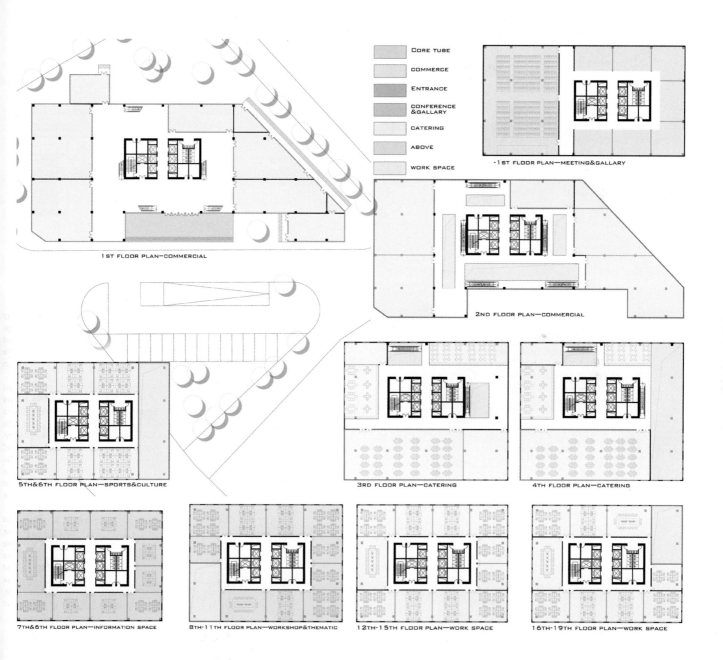

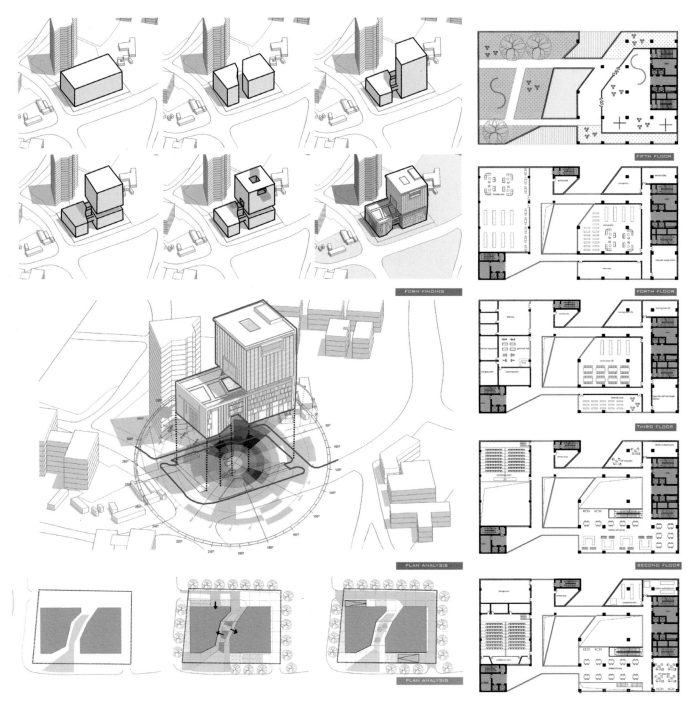

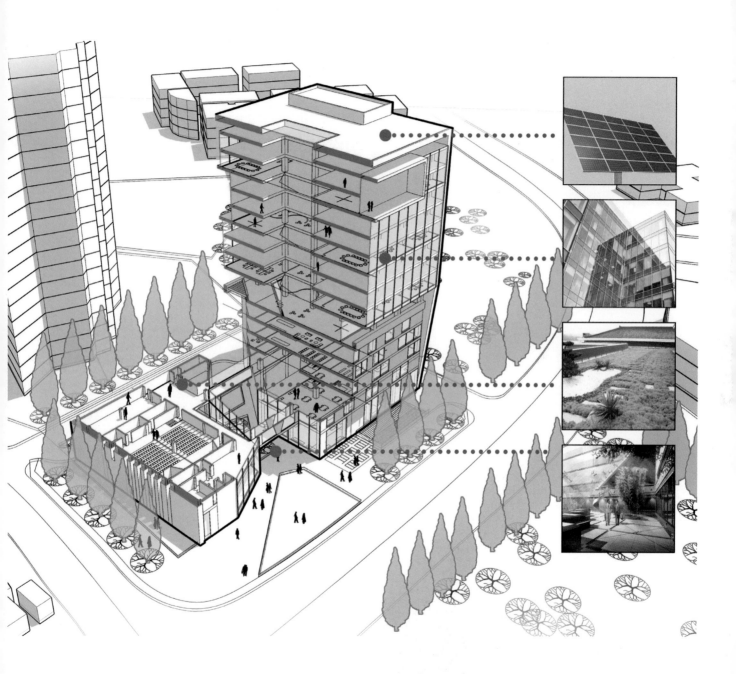

建筑设计研究（二）DESIGN STUDIO 2
多层木结构建筑设计
DESIGN OF A MULTI-STOREY TIMBER STRUCTURE BUILDING
乌都·特尼森 孟宪川

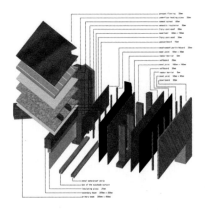

近年来，木材作为建筑材料变得越来越重要。联合国宣布2011年为"国际森林年"，这种全球生态意识的推动也激发了建筑行业的新思维方式。防火和隔音方面的改进以及计算机辅助微积分在制造中的使用，大大提高了木材的设计潜力和建筑应用。最近对建筑规范的修改允许在大城市地区建造木制多层建筑。因此，木材作为最古老的建筑材料之一，与"现代"材料——钢和混凝土相比不再处于不利地位；我们现在完全可以看到木材在建设未来项目中发挥重要作用的潜力。

本次工作营专注于新大都市木制建筑概念。在中国丰富的木结构建筑历史背景下，我们将借助于多层建筑，探索传统和创新技术如何在城市环境中创造新形式的木结构建筑。办公楼的建筑类型将有助于体验木制建筑的不同原则、结构和体积的关系以及一般和特定空间的作用。

本次工作营将研究不同的传统、最佳实践和推测性建筑原则，从单个接头或单元开始，演变成多层建筑系统。建筑如何通过单一元素来表达？将产生哪些空间后果？在第二步中，将介绍方案和选址。在每周简短的讲座中，将为学生们提供材料特性、结构性能、制造、防火和防噪方面的基本知识。在设计过程中，为理解试验过程、材料和形式之间的相互关系，物理模型将发挥重要作用。

In recent years, wood has become increasingly important as a building material. The UN declared the year 2011 the "International Year of Forests" and this global drive for ecological awareness also inspired a new way of thinking within the building sector.Improvements in fire protection and acoustic insulation as well as the use of computer-aided calculus in manufacturing have greatly enhanced the design potential and architectural application of wood. Recent adaptions of building codes allow the erection of wooden multi-storey buildings in metropolitan areas. Thus wood, as one of the oldest building materials, is no longer disadvantaged in comparison with the "modern" materials—steel and concrete; we can now see the potential of wood to play a vital role in building projects of the future.

This studio will focus on the concepts for a new metropolitan timber architecture. Against the background of the rich history of timber construction in China, also in multi-storey buildings, shall be explored how traditional and innovative techniques could create new forms of wood constructions in urban context.The building type of the office building will serve to experience different principles of timber construction, the relation of structure and volume, and the role of generic and specific spaces.

Different traditional, best practice and speculative construction principles will be studied, starting from a single joint or unit and evolving to a multi-storey construction system. How will the architectural expression be informed by the single element and which spatial consequences will arise? In the second step the program and the site will be introduced. In short weekly lectures, the basic knowledge of material properties, structural performance, manufacturing, fire and acoustic protection will be provided.During the design course, physical models will play a substantial role in order to understand the interrelation of the experiment process, material and form.

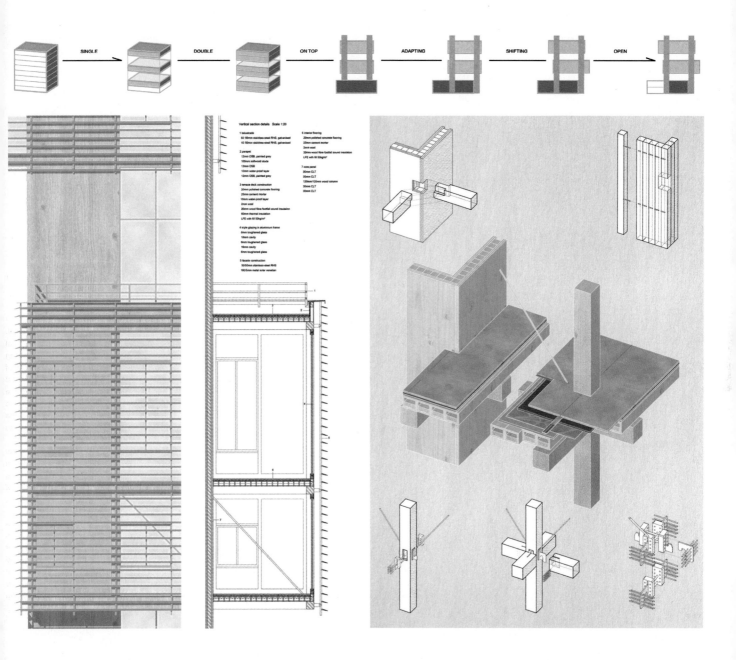

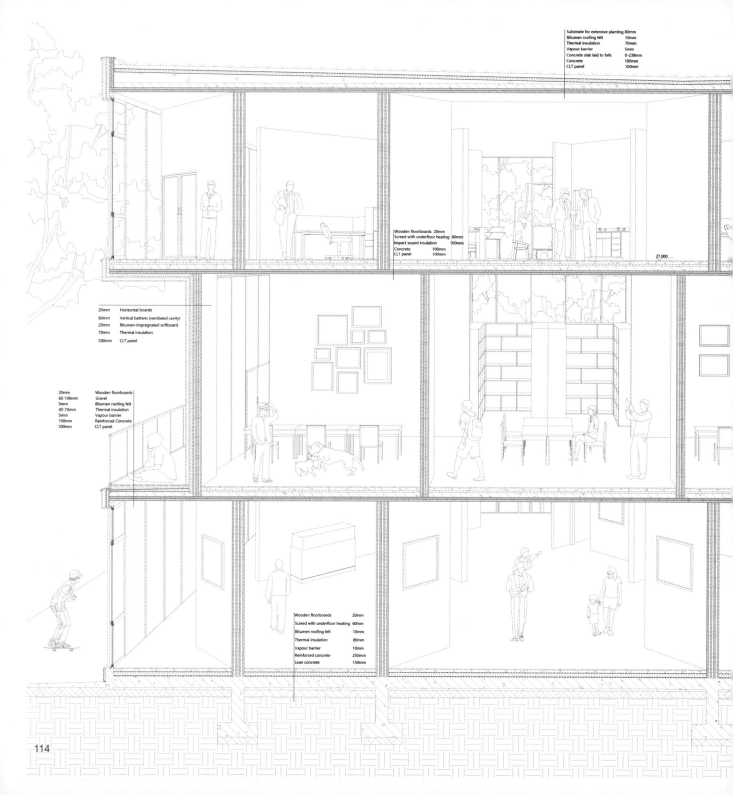

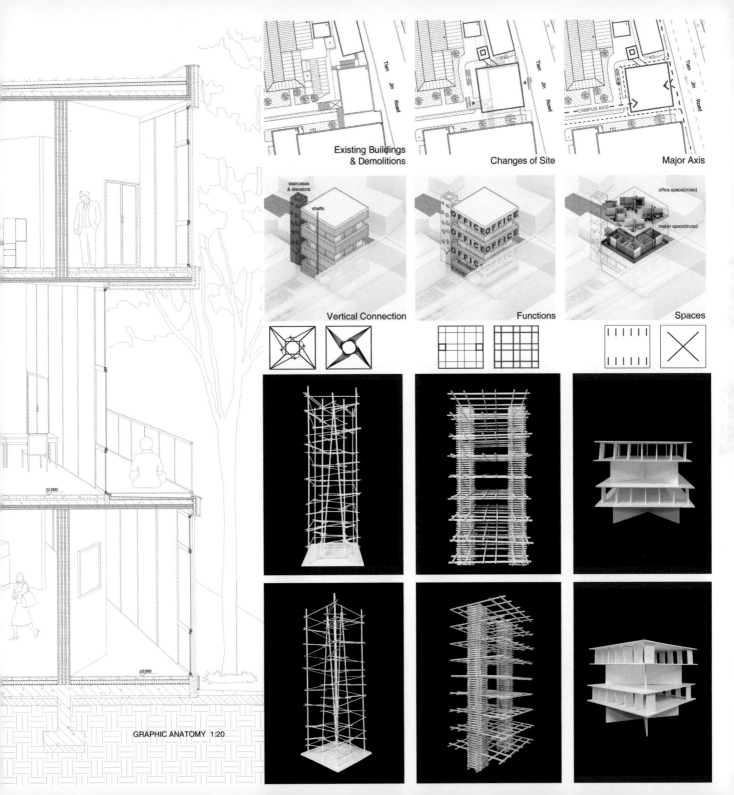

Load Transfer

The stucture of the original building

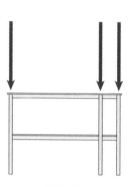

The load path

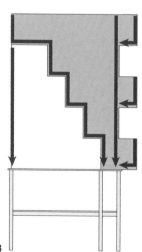

The load path of the wood structure

Construction System

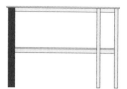

Bold columns

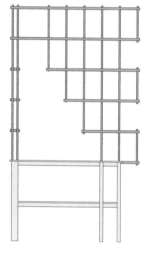

Add columns and beams

Add the braces

Process Models

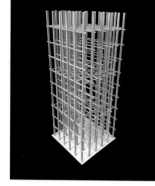

The first model
Concept: forest
Construction principle: columns and beams

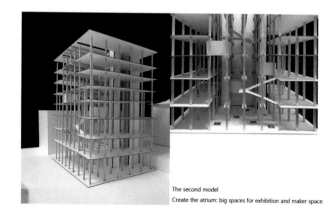

The second model
Create the atrium: big spaces for exhibition and maker space

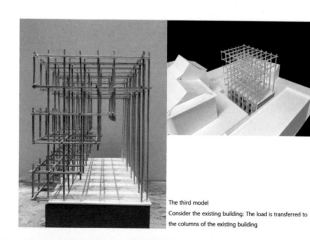

The third model
Consider the existing building: The load is transferred to the columns of the existing building

建筑设计研究（三）DESIGN STUDIO 3

异质类型：建筑、基础设施和地景（1）相地
HETEROTYPE: ARCHITECTURE, INFRASTRUCTURE, LANDSCAPE(1) SITE DESCRIPTION

卡里·西雷斯 丁沃沃

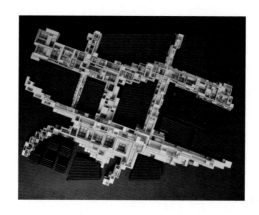

本课程旨在介入21世纪的设计想象，提出了5个对建筑实践未来可能产生影响的关键问题：
（1）除了那些已经沿着新自由主义资本家想象的方式设想的方案之外，我们能梦想出设计的替代未来方案吗？
（2）我们能梦想超出纯粹经济或技术的设计新前景吗？
（3）我们能否梦想通过提出新的共同目标来加强更具包容性的公民领域，从而使所设计的环境更好地服务所有居民？
（4）我们能否梦想更好的方式，通过加入混合设计能力，鼓励社会持续参与集体生活？
（5）我们能否梦想从我们创造的建筑、基础设施和景观中为世界做出更有益的贡献？
设计工作室将探索建筑造型、基础设施服务格式化和景观地域构成之间的混合"中间地带"。通过在南京市内的一系列相互关联的现场和工作室实习，学生们将为各自的城市环境开发出新的方案，将单一规划的建筑、未充分利用的基础设施和被忽视的景观场所改造成多模式的社会连接体。
研讨会讲座以《镜像影响：领土的故事》一书（马克·安吉利尔，卡里·西雷斯，Ruby Press，柏林，2018）的研究为基础。该书探讨了新自由主义资本主义的多重地缘经济理性如何被转化到不同世界地区的社会和物质领域。每周一次的讲座将探讨在特定世界背景下的这一主题。

Aiming to intervene on the 21st century design imaginary, the course poses 5 critical questions that bear on possible futures of architectural practice:
(1) Can we dream of alternative future scenarios for design than those already envisioned along the lines of a neoliberal capitalist imaginary?
(2) Can we dream of new prospects for design beyond the purely economic or technical?
(3) Can we dream of ways to make the designed environment work better for all inhabitants by initiating new common purposes that would reinforce a more inclusive civic realm?
(4) Can we dream of better ways to encourage sustained social engagement in collective life by joining hybrid capacities of design?
(5) Can we dream of more beneficial contributions to the world from the architecture, infrastructure, and landscapes that we produce?
The design studio will explore a hybrid "middle ground" between architectural form-making, infrastructural service formatting, and landscape territorial composing. Through a series of interrelated field and studio exercises within the city of Nanjing, students will develop scenarios for adapting mono-programmed architecture as well as underutilized infrastructure and neglected landscape sites into multi-modal social connectors for their respective urban settings.
The seminar lectures are based on research that was published the book *Mirroring Effects: Tales of Territory* (Marc Angélil / Cary Siress, Ruby Press, Berlin, 2018). The book examines how multiple geo-economic rationalities of neoliberal capitalism are translated into social and physical territories in different world localities. Each weekly lecture will explore this theme which is a specific world setting.

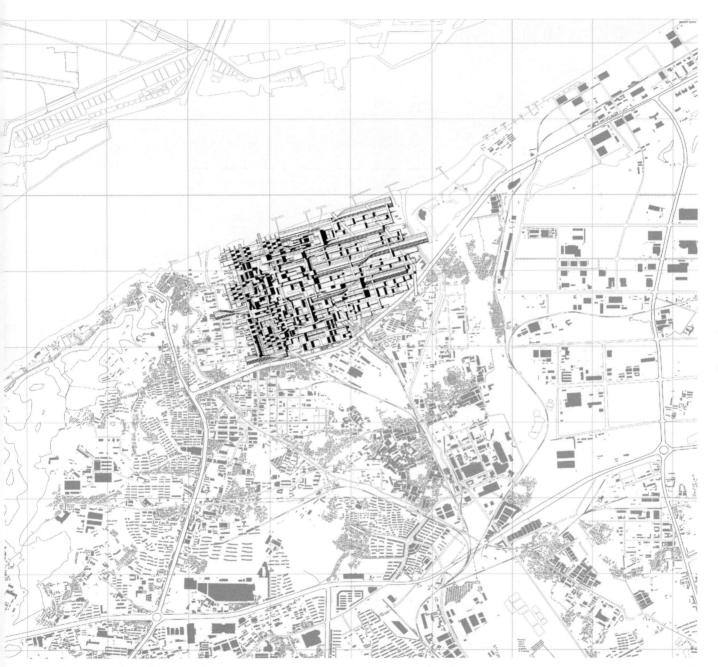

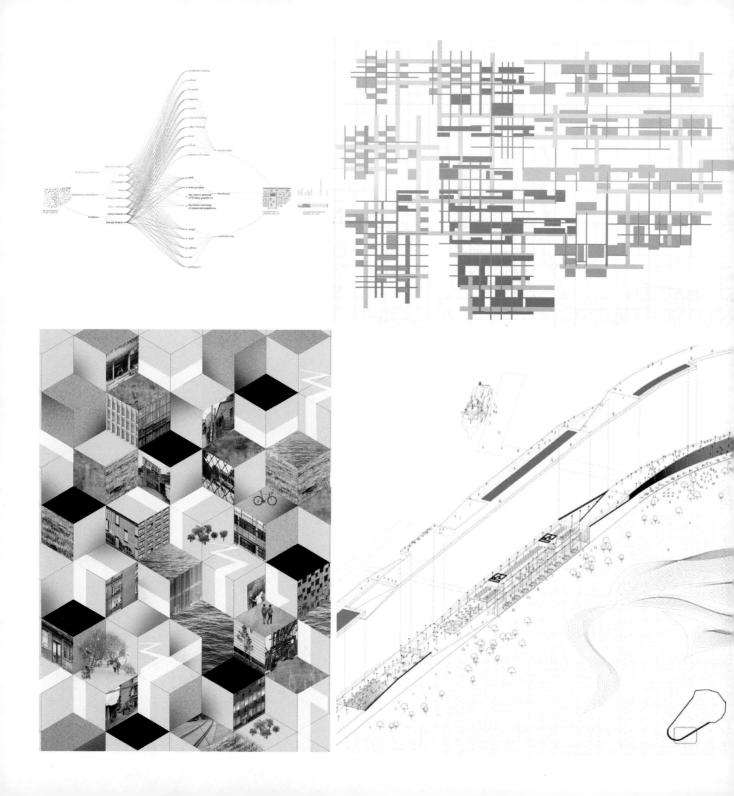

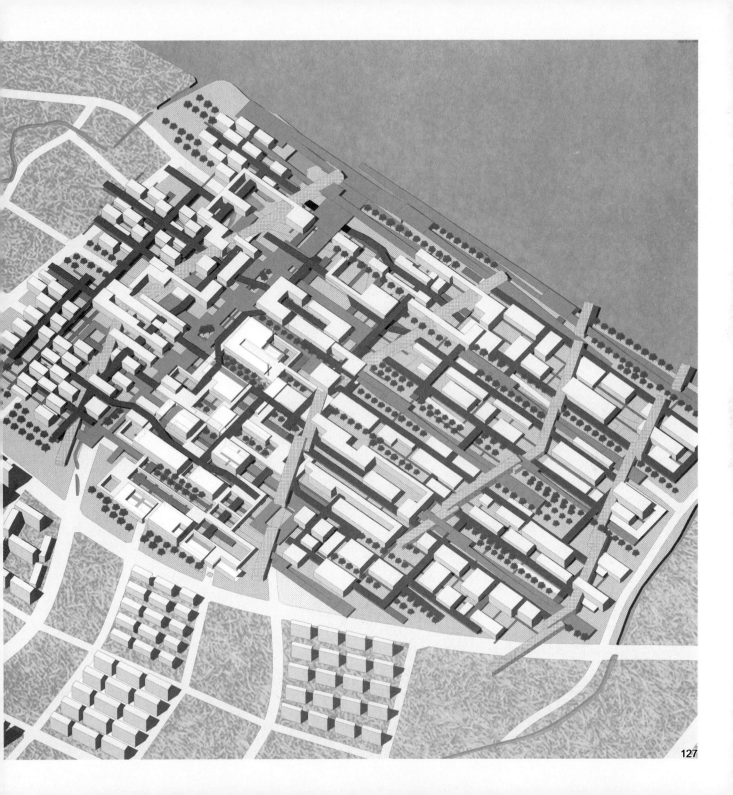

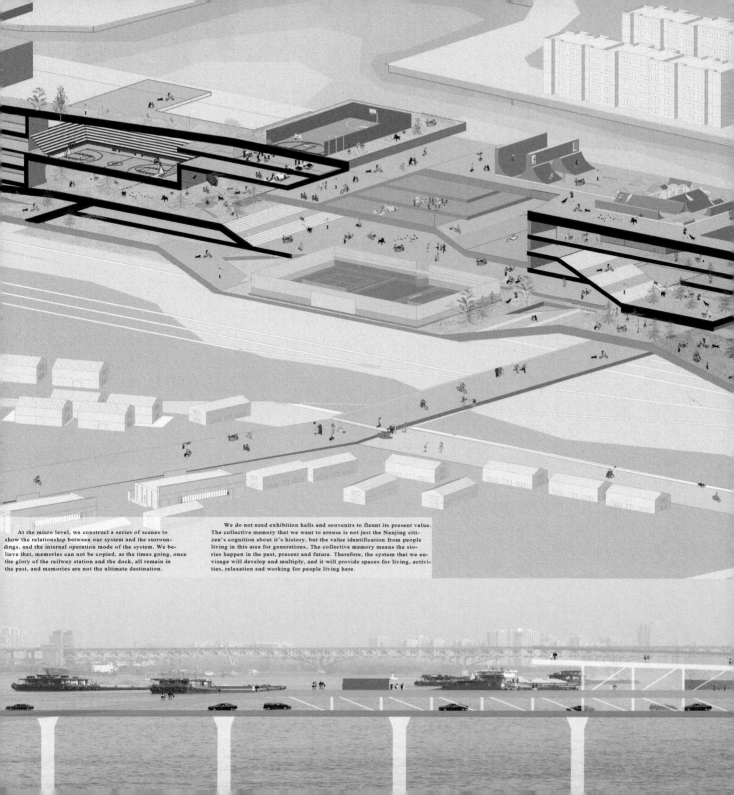

At the micro level, we construct a series of scenes to show the relationship between our system and the surroundings, and the internal operation mode of the system. We believe that, memories can not be copied, as the times going, once the glory of the railway station and the dock, all remain in the past, and memories are not the ultimate destination.

We do not need exhibition halls and souvenirs to flaunt its present value. The collective memory that we want to arouse is not just the Nanjing citizen's cognition about it's history, but the value identification from people living in this area for generations, The collective memory means the stories happen in the past, present and future. Therefore, the system that we envisage will develop and multiply, and it will provide spaces for living, activities, relaxation and working for people living here.

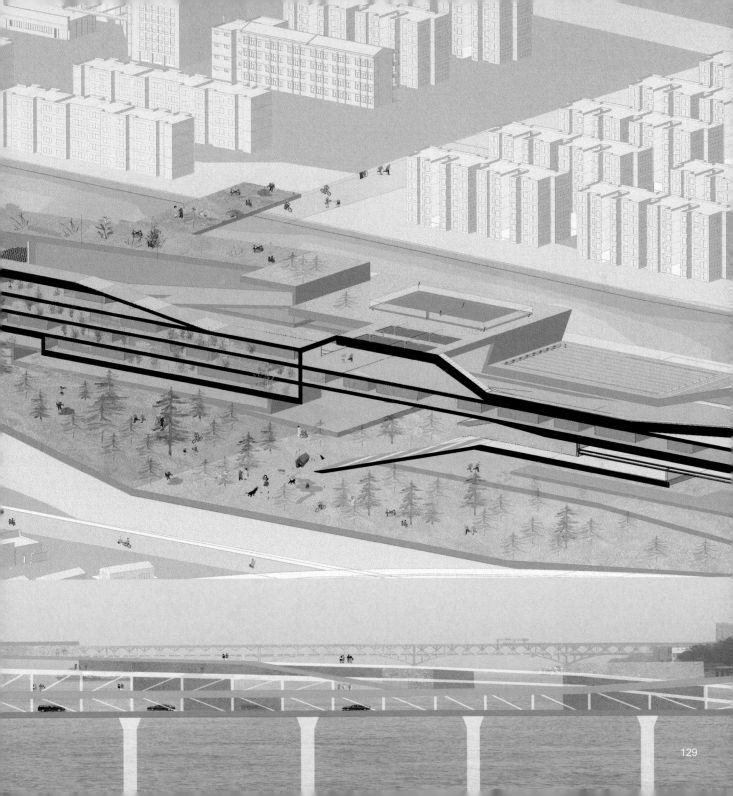

建筑理论课程 ARCHITECTURAL THEORY COURSES

"建筑史方法"与李渔的《十二楼》
"ARCHITECTURAL HISTORY METHODOLOGY" AND LI YU'S *TWELVE PAVILIONS*

胡恒

"建筑史方法"是一门研究生的历史课。顾名思义，它以讲解历史研究的方法为主旨。从2014年开始，我琢磨着摆脱掉这门课从文本到文本的固有循环，将知识学习和创作活动结合起来。也就是说，将历史、文字、图像与设计结合起来。我认为，硕士阶段的学生有能力、更有必要来进行这种综合型的工作。所以，这门课的主干部分是讲课，而更重要的部分是作业。因为这一多层面的综合工作需要作业才能得以实现。

作业的构想是这样的。以某一小说作为基本文本，让学生将小说的文字转化为一项设计。也即，从文学作品到建筑与空间设计。我选取的小说是清初作家李渔的《十二楼》。这部小说有十二章，一章一个独立的小故事。每个故事以"楼"为题，分别是合影楼、夺锦楼、三与楼、夏宜楼、归正楼、萃雅楼、拂云楼、十卺楼、鹤归楼、奉先楼、生我楼、闻过楼。这些故事轻松诙谐，是李渔的一贯风格。每个故事各有明确的历史背景——从宋到明末。内容多属才子佳人及文人雅事，也有悲欢罹难、市井趣闻。虚构与写实交织在一起，言浅意深，内含的格局很大。

作为建筑教学的基础文本，《十二楼》这本小说有几个先天的优势。一则，十二个故事都与特定的建筑或空间"楼"有着密切关联，具有原生的"建筑性"。二则，故事不长，且都涉及具体的历史、地理语境，细节充沛，学生可从历史、社会、文学、建筑等不同方面来阅读理解。三则，十二个故事方便学生分组，容易操作。

最初的作业设置是在零经验的情况下进行的，我将其当作一次实验。作业的要求也很简单，两条思路：一个是还原小说中的建筑原貌；一个是设计一个建筑，来讲解这个故事。同学们可以任选其一加以深入。

第一个思路是我的首选意向，多年前我在看这本小说的时候，就对其中的建筑空间很是好奇。十二"楼"非常多样，地域既涵盖大江南北，类型也是花样百出——有私家园林，也有皇宫内府；有寻常小院，也有华宅精舍；有沿街店铺，也有贡院佛寺。有些"楼"还有超出建筑之外的空间戏剧性。比如第一篇"合影楼"，故事发生在一个大宅院里。因为家族内部的矛盾，一家分作两户，其中一个家主修了一道"高墙"从内部将大宅院强行分开，表示再不往来。这道墙延伸到后院的水池上，它架在水面，把水池也虚拟分成两半。这个场景有点超现实的味道。如果说第一个思路是考证与还原，第二个思路则是概念与设计。将文学作品转化成视觉艺术很常见，比如电影、戏剧、绘画。相比之下，文学转化为建筑较为少见，建筑史上只有特拉尼的但丁纪念堂设计方案属于此类。总体来说，这是一种比较"偏门"、有难度的创作。在这个作业里，我没有交代太多的框框和要求，打算先看看大家的思路能够打开到哪种程度，发散出多少可能性。

第一次的作业收上来。一大半是做历史还原，只有很少几组选择设计。以"合影楼"为例，有两份作业分别采用了还原与设计两条思路。小说发生在元朝

《合影楼》插图

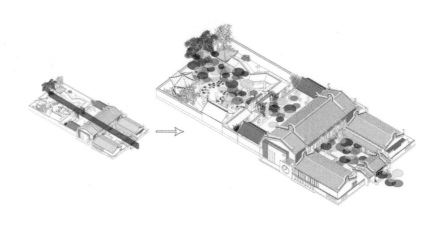

合影楼设计　The Design of Heying Pavilion

至正年间广东韶州府曲江县的两户缙绅之宅中。"还原"小组从年代、地域着手，研究了元代岭南的私家花园与宅院的各项特征，作业的最后结果是一个与小说故事相近规模的宅园方案。我称这类"还原"小组为"考据派"，其作业也像一篇历史方向的学术论文。设计方向的小组做了一部分考据工作，并进一步把"历史"建筑进行现代语言的转化，我称这种小组为"设计流"。从历史知识考证到现代空间设计这一步并不容易跨过来，不过一旦转换过来，大家就会发现，以现代建筑语言重建"十二楼"其实相当有趣。小说中能够起到转秘作用的元素有很多：角色的性格、人物关系、建筑特点、特定器物、历史场地等等。

2015年的作业发生重大变化。同学们几乎全部选择了第二个思路，无人再做学术"考据派"，都成了"设计流"。大家都离开了前一年的用现代"十二楼"置换古典"十二楼"的做法，把"设计流"向前推进了一大步。古典叙事的现代转译出现了三种新模式。第一个是抽取故事的结构，加以现代空间化表现。这个结构可以是故事的主结构（《十二楼》中有几个故事的主结构有着非常明显的抽象形式），也可以是同学们在阅读过程中发掘出来的隐秘结构。第二个是将故事的主场景的空间特征进行专业整理，再将其在当下的现实世界中找到对应，使之变形为一个现实中新的建筑任务，比如把"合影楼"变成游泳馆的更衣室。第三个是对故事的主场景进行现代空间语言的处理，并将形成的空间体道具化，再推敲它在现实世界的存在可能。

三种模式都实现了以现代空间叙述古代故事的目的。一个是把叙事结构进行空间化，用新的结构来重现故事；一个是用现代空间语言来讲述一个新故事；一个是将场景转化成现代空间装置，再安置在城市里使之发生作用。三种模式中叙事结构的空间化最多，同学们对故事结构的分析找到很多空间原型与之对应，比如迷宫、回文、艾舍尔的幻觉画、圆形监狱、葫芦、莫比乌斯环、盒中盒等等。新的建筑功能置换也有几例很精彩，比如公共浴室、餐厅、剧院。这批作业要求是一套文本，我发现有一些设计如果做成模型会更有趣。所以，我对下一年的作业补充一个要求：必须做模型，而且底板用实木，且尺寸统一。

2016年的作业比前一年又有进步。大家对古典文本的现代转译的各种常规路数已大体熟悉，开始探索新的可能性。一方面，同学们对"楼"的故事隐含着人生态度、世界观开始有所领悟。另一方面，大家从故事中提取的概念要素越来越有趣，比如"合影楼"中的"合"与"影"，"归正楼"中的"正"字，"鹤归楼"中某配角宋徽宗的名作《瑞鹤图》。这些都成了设计的切入点，空间叙事开始概念化，设计不再是知识与技术操作，而是对某种形而上之物的阐述。再者，在不同知识领域的跳跃也比前两年要灵活得多。电影、戏剧、电子游戏、艺术史、建筑史、名人等等均有涉足。模型作为最终表现形式，这一要求被证明很有效，它在使得设计更为切实之外，也让不同作业的优劣差异十分明朗。

对于2017年的作业，我的要求明确为五点。第一，分析故事结构、情节特征、人物关系、历史背景，并整理出一个独特的结构。第二，分析故事中涉及的建筑、空间的形态,梳理带有空间属性的道具。第三,寻找阅读故事后的共鸣点,感知故事中的特定气息。第四,搜寻共鸣点所可能具有的形式原型。第五,将以上几点综合起来形成一个"设计"。

这期作业在三个方面出现新的动向。其一,同学们对故事的气质、气息的捕捉有所成效。前几年,大家对故事的解读集中在剧情结构、人生哲理、价值观上。现在开始提炼更为微妙的东西。其二,大家对自己与故事的共鸣点逐渐看重,设计的切入口往往由此而来。共鸣点是我在这一年特别强调的,我希望同学们能找到共鸣点与个人经验的关系,个人经验的介入会让设计的个人化色彩变得强烈起来。

之前的模型要求只是底板统一,这一年我提了几项新要求。第一个,模型材料可以尝试将几种硬质、软质差异较大的类型做到一起。第二个,我希望有些模型能够动起来,不一定都做成建筑模型那种静态。有几个模型达到了我的要求。比如"萃雅楼",模型是一个圆锥体顶着一块圆板,圆板边沿和锥体底部有皮筋和鱼线拉住。圆板上放着一个球,如果把某根皮筋剪断,圆板就会侧翻,球掉下圆板。在"奉先楼"

里,同学们斥"巨资"购买了一个漂亮的古色古香的铁盘,在上面用棉花、宣纸、盐搭建了一个小小的空间体,然后点火将之烧却,在盘中央留下一堆灰烬。模型向装置转型是这一年作业的显著特点。"设计作业"出现了一点"当代艺术"的感觉。

回顾四年的作业,其变化的幅度超出我本来的预期。从"考据派"到"设计流",还不期然地跨到"当代艺术"。就设计来说,无论是故事的解读,还是概念的提取、形式的设计、模型的制作,逐年都有进步。另一方面,四年的作业只能说是打下了一个基础,各个环节都还存在很多问题。比如说故事的结构分析,现在出现的模式已有不少,但仍有待变换角度深度挖掘。某种程度上,第一步的分析会影响后续工作的成效,而更高层面的工作——对故事的分析、对形式的分析、对自我的分析三位一体——现在只是初现端倪,要想融会贯通,还有很长的路要走。

就作业的成果来说,我发现《十二楼》中有几个楼比较容易出好设计,比如合影楼、萃雅楼、拂云楼。与此同时,有几个楼一直都做得不理想。比如夺锦楼、生我楼、闻过楼。其中原因我尚在思考,希望在以后的作业里能够逐步把它们"解决"。

"Architectural history methodology" is a postgraduate history class. As its name implies, it aims to lecture methodology of historical research. Since 2014 I have pondered to get rid of the inherent cycle from text to text in this class, and combined knowledge learning with creative activity. In another word, I combine history, text, image with design. I believe that postgraduate has the ability and necessity to have such comprehensive work. Therefore, the main part of this class is lecture, but the more important part is homework, for the comprehensive work is reflected by homework.

Here is the concept of homework: I take some novel as the basic text, and make students transform the text of novel into a design. That is, from literary work to architecture and space design. I choose *Twelve Pavilions* written by Li Yu, a writer in early Qing Dynasty. This novel includes twelve chapters, and every chapter has a separate story. Every story is themed in "pavilion", respectively Heying Pavilion, Duojin Pavilion, Sanyu Pavilion, Xiayi Pavilion, Guizheng Pavilion, Cuiya Pavilion, Fuyun Pavilion, Shijin Pavilion, Hegui Pavilion, Fengxian Pavilion, Shengwo Pavilion, Wenguo Pavilion. These stories are relaxed and humorous, in the style of Li Yu as he always was. Every story has a clear historical background—from Song Dynasty to the late Ming Dynasty. The content is mainly about literary pursuits of gifted scholars and beautiful ladies, also including joys and sorrows, folk funs. Fiction is interwoven with reality, with plain language but profound meaning and connotation.

As the basic text of architectural teaching, *Twelve Pavilions* has several congenial advantages. Firstly, the twelve stories are closely related to specific architecture or space "pavilion", with native "architecture". Secondly, the story is not long, but involves specific historical, geographic context with abundant details, students can read and understand from history, society, literary, architecture. Thirdly, the twelve stories are convenient to group students, and easy to operate.

The initial homework was assigned without experience. It was an experiment to me. The homework requirement was simple. There were two ideas: one was to restore the original appearance of architecture in the novel; the other was to design an architecture to tell this story. Students could choose any one to researched thoroughly.

The first idea was my preferred intention. When I read this novel years ago, I was curious about its architectural space. Twelve "pavilions" were diversified, with region covering both sides of Yangtze River with various types: there was private garden and imperial houses in palace; there was common courtyard and luxury mansion; there was store along the street and examination hall and temple. Some "pavilion" had space drama beyond architecture. For example, the first chapter Heying Pavilion told a story in a mansion. A family was divided into two households due to contradiction inside the family, in which one householder built a "high wall" to separate the mansion from inside by force, which implied they would have no contact ever. This wall was extended to the pool in the backyard. It was erected above the water surface, and virtually halved the pool. This scene was somewhat surreal. If the first idea was verifying and restoring, the second idea would be concept and design. It is common to transform literary work into visual art, such as film, drama and painting. By contrast, it is rare to transform literary into architecture. In the history of architecture, only Terragni's design scheme of Danteum is of such kind. Generally speaking, this is a "less popular" and difficult creation. In this homework, I did not set too many restrictions or requirements, I wanted to see how open everybody's mind could be and how many possibility they would show.

After collecting the homework, I found that most focused on history restoration, only few groups chose design. Heying Lou was taken for an example, two work respectively adopted the idea of restoration and design. The story in the novel happened in two mansions of retired government officials in Qujiang County, Shaozhou Prefecture, Guangdong during the Zhizheng year, Yuan Dynasty. The "restoration" team started with time and region, and researched the characteristics of private garden and mansion in the south of the Five Ridges in Yuan Dynasty. The final result of the homework was a mansion plan in similar scale with the story in the novel. I called such "restoration" team as "textual research school". Their homework was like an academic historical thesis. Team that focused on design conducted a part of textural research work, and further transformed "historical" architecture into modern language. I called such team as "design genre". It is difficult to transform from historical knowledge textual research to modern space design. However, once being transformed successfully, everybody will find that it is interesting to rebuild "twelve pavilions" by modern architectural language. There are many elements that play a role of transforming in the novel: personality of character, relationship of character, architectural characteristics, specific utensils, historical site, etc.

The works in 2015 had significant change. Almost all students chose the second idea and became "design genre". Instead of replacing classic "twelve pavilions" by modern "twelve pavilions", everybody greatly promoted "design *genre*" forward. Three new modes appeared in modern translation of classic narration. The first mode extracted the structure of story and expressed it in modern space. This structure could be the main structure of story (the main structure of some stories in *Twelve Pavilions* has very obvious abstract form), or the implied structure dug by students during reading. The second mode professionally arranged the space characteristics of main scene of story, found relevant correspondence in the real world, and made it a new architectural task in reality, such as turning "Heying Pavilion" into the locker room in natatorium. The third mode processed the main scene of story into modern space language, made the formed space property, then reckoned the possibility for it to exist in the real world.

The three modes realized the purpose of telling ancient story by modern space. One mode made the narration structure a space, and reproduced the story in new structure; One mode used modern space language to tell a new story; One mode turned scene into modern space device, then settled it in city to function. Among the three modes, most homework made the narration structure space. Students found many space prototypes to correspond with the story structure, such as labyrinth, plaindrome, illusion painting of Escher, panopticon, cucurbit, Mobius band, box in box, etc.. There were also some excellent cases of new architectural function replacement, such as bathhouse, dining hall, theater. The works requirement was a set of text. I found it would be more interesting to model some design. Therefore, I proposed a supplemental requirement on the homework of next year: model is required, and the base plate shall adopt solid wood in unified dimension.

The works in 2016 made further progress. Everybody was familiar with the routine for modern translation of classic text, and began to explore new possibility. On one hand, students began to comprehend the attitude towards life, world view implied in the story of "pavilion". On the other hand, the conceptual elements drawn by everybody from the story became more and more interesting, such as "he" and "ying" in "Heying Pavilion", "zheng" in "Guizheng Pavilion", Crane Painting, the famous work of Emperor Huizong of Song Dynasty, as some supporting character in "Hegui Pavilion", all became the entry point of design. Space narration became conceptualized. Design was not only knowledge and skill operation, but the narration of some metaphysical object. The leap among different knowledge fields was more flexible, involving film, drama, video game, art history, architecture history, celebrity, etc.. The requirement on using model as the final pattern of manifestation was proved effective. Besides making the design more realistic, it also made it clear to see the quality of works.

I proposed five requirements on the homework in 2017. Firstly, to analyze story structure, plot characteristics, character relationship, historical background, and arrange a unique structure. Secondly, to analyze the form of architecture, space involved in the story, arrange properties with space attribute. Thirdly, to find out the resonance after reading story, feel the specific breath in the story. Fourthly, to find the possible form prototype of resonance. Fifthly, to summarize aforesaid points and form a "design".

New trend appeared on three aspects in the homework. Firstly, students had effect in capturing temperament and breath of story. Few years ago, they focused on plot structure, life philosophy and value in the story. Now they began to distill finer things. Secondly, they attached more importance to the resonance with the story, which opened the approach of design. Resonance was what I emphasized in this year. I hoped that students could find the relation between resonance and personal experience. The intervention of personal experience would make personal color stronger.

The previous model requirement was unified base plate. In this year I proposed some new requirements. Firstly, to combine hard and soft model materials. Secondly, I hoped that some model could be movable, not as static as the building model. Some models met my requirement. For example, the model of "Cuiya Pavilion" was a cone with a disk on top, the edge of disk and bottom of cone were pulled by rubber band and fish wire. A ball was placed on the disk, if some rubber band was cut, the disk would roll over, and the ball would fall off the disk. In "Fengxian Pavilion", students spent "huge fund" buying a beautiful antique iron disk, built a small space body by cotton, rice paper, salt above, then burnt it and left ashes in the center of the disk. The transformation from model to device is the obvious characteristics of the homework in this year. "Design homework" shows the sense of "contemporary art" somewhat.

Reviewing the homework in the four years, I find that the change is beyond my expectation. From "textural research school" to "design genre", and unexpected leap to "contemporary art". For design, no matter it is the understanding of story, or extraction of concept, design of form, making of model, progress is made every year. On the other hand, the homework in four years only lays a foundation, there are still many problems in all links, such as story structure analysis, despite of many modes, and it is still be to further dug from different perspectives. To some extent, the analysis in the first step will affect the effect of subsequent work. Work on higher level—trinity of analysis of story, analysis of form, analysis of self—is just over the horizon. It is a long way to go to achieve mastery through a comprehensive study.

On the achievement of homework, I find it easier to achieve good design in several pavilions, such as Heying Pavilion, Cuiya Pavilion, Fuyun Pavilion. At the same time, design of some pavilions is not ideal, such as Duojin Pavilion, Shengwo Pavilion, Wenguo Pavilion. I am still thinking about the reason, and hope that they can be "solved" in the homework in the future.

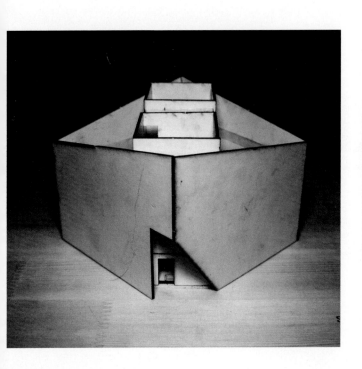

合影楼设计 The Design of Heying Pavilion

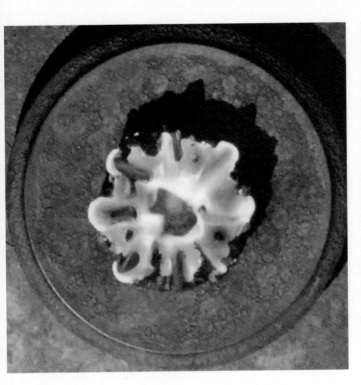 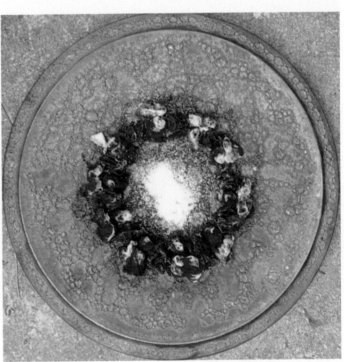

奉先楼设计　The Design of Fengxian Pavilion

圆形作为母题

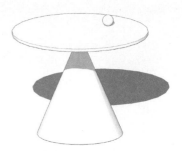
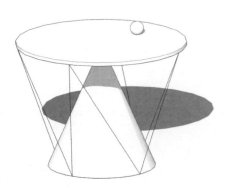
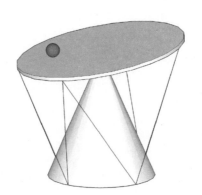

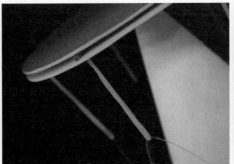
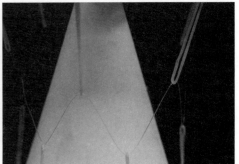

萃雅楼设计　The Design of Cuiya Pavilion

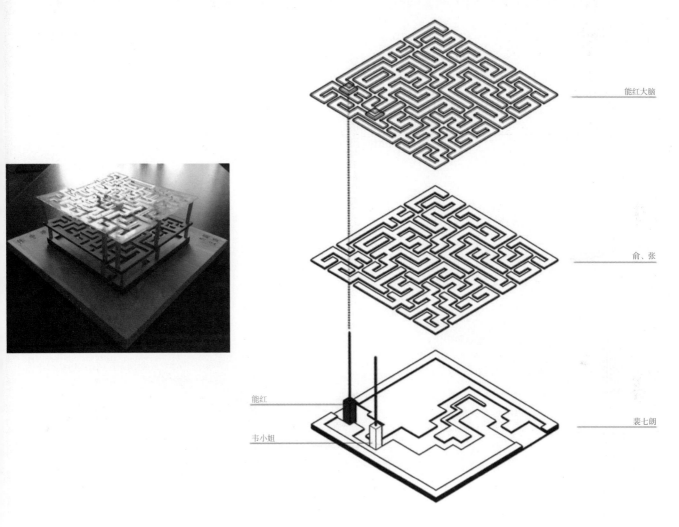

能红大脑

俞、张

能红

韦小姐

裴七朗

拂云楼设计　The Design of Fuyun Pavilion

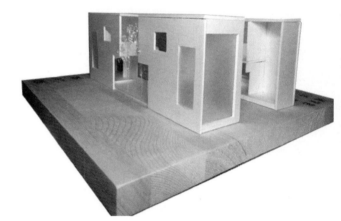

正 —— 止 —— 正

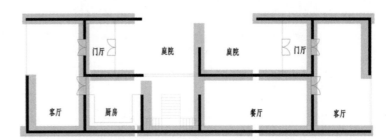

正 —— 止 —— 正

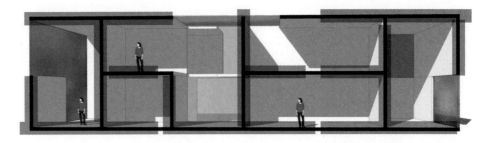

归正楼设计 The Design of Guizheng Pavilion

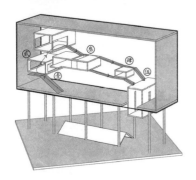
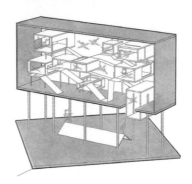
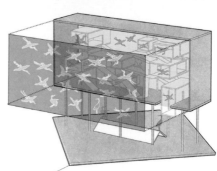

鹤归楼设计　The Design of Hegui Pavilion

建筑设计课程
ARCHITECTURAL DESIGN COURSES

本科一年级
科学与艺术
· 季鹏
课程类型：必修
学时学分：32学时／2学分

Undergraduate Program 1st Year
SCIENCE AND ART · JI Peng
Type: Required Course
Study Period and Credits: 32 hours/2 credits

教学目标
通过讲课和艺术实践并举、重在实践的教学路径，提高理工科学生的设计意识。同时通过艺术实践，开拓理工科学生的眼界，提升自身内在综合素养，为未来的研究与实践储备创意创新的潜力。
授课方式
导论为讲课，艺术实践分为10个工作室，每名学生选修3个工作室。
工作室实践内容
包括绘画（2个）、摄影、数字媒体艺术（2个）、图形与丝网版画、陶艺、蜡染、布艺、模型。

Training Objective
Through the teaching path that takes importance to both teaching and artistic practice, and focuses on practice, improve the design consciousness of students of science and engineering. At the same time, through artistic practice, open up mind of students of science and engineering, improve internal comprehensive quality, and reserve creative and innovative potential for future research and practice.
Teaching Method
Teaching for introduction, 10 studios for artistic practice, each student selects 3 studios.
Content of Studio Practice
Painting (2), photography, digital media art (2), graphic and silk screen print, pottery, wax printing, fabric art, model.

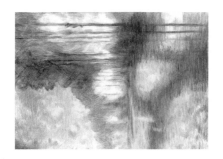

本科一年级
设计基础（二）
· 鲁安东　唐莲　尹航　孟宪川　黄华青
课程类型：必修
学时学分：64学时／2学分

Undergraduate Program 1st Year
BASIC DESIGN 2 · LU Andong, TANG Lian, YIN Hang, MENG Xianchuan, HUANG Huaqing
Type: Required Course
Study Period and Credits: 64 hours/2 credits

教学内容
采用课内多模块选修制度。整个课程分为3个阶段，每个阶段时长5周。分别为感知（转化能力训练）、分析（制图能力训练）、创造（动手能力训练）。
学生分成A班（小班）和B班（大班）。A班每班20人，共2个班，采用设计教学（studio）的动手实操形式。B班1个班60~80人，采用大课授课（lecture）形式。A班共有6个教学模块（每阶段2个），A班学生在学期期间共选修其中3个模块。6个模块为不同的设计练习，具体内容由教师进行设计。
教学要点
A类教学模块教学主要内容包括：
1. 感知：培养感受与思维的协同，包括对空间的感受、对身体的感受、对氛围的感受等。
2. 表达：培养表达与思维的协同，掌握不同媒介、可视化手段，让学生具备运用表达手段进行想象和思考的能力。
B类模块内容为建筑学介绍、建筑鉴赏、建筑历史等带有通识性质的讲课。

Teaching Content
A multiple-module elective course system in class is implemented. The entire course is divided into 3 stages, each of which lasts 5 weeks, respectively sensing (transformation capacity training), analyzing (drawing capacity training), creating (handwork capacity training). The students are assigned into Class A and Class B. Class A has 2 classes are arranged with each of which containing 20 pupils. A studio classroom featured with interactive engagement and active learning is adopted in Class A. Class B, which has a capacity of 60~80 people, adopts lectures as a teaching method. Class A has 6 teaching modules (2 for each phase. Among which 3 of them are required by pupils of Class A as elective courses during the semester. The 6 modules are studio practices distinct from each other, and specific contents of these modules are to be designed by the instructor.
Teaching Essential
Main contents of class A teaching module include:
1. Sensing: Cultivate the synergy between felling and thinking, including the feeling of space, the feeling of body, the feeling of atmosphere.
2. Expressing: Cultivate the synergy between expression and thinking, and the ability to manage different media and visualization means, empowering the students have ability of imaging and thinking by expressive means.
The content of class B module includes introduction to architecture, architecture appreciation, architectural history and other lecture with general education nature.

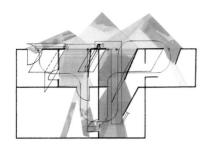

本科二年级

建筑设计基础

· 刘铨 冷天

课程类型：必修

学时学分：64学时 / 4学分

Undergraduate Program 2nd Year
ARCHITECTURAL BASIC DESIGN · LIU Quan, LENG Tian
Type: Required Course
Study Period and Credits: 64 hours/4 credits

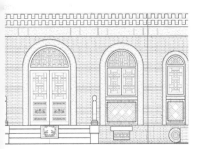

课题内容
　　认知与表达
教学目标
　　本课程是建筑学专业本科生的专业通识基础课程。本课程的任务主要是一方面让新生从专业的角度认知与实体建筑相关的基本知识，如主要建筑构件与材料、基本构造原理、空间尺度、建筑环境等知识；另一方面通过学习运用建筑学的专业表达方法来更好地掌握这些建筑基本知识，为今后深入的专业学习奠定基础。
教学内容
　　1.认知建筑
　　（1）立面局部测绘
　　（2）建筑平、剖面测绘
　　（3）建筑构件测绘
　　2.认知图示
　　（1）单体建筑图示认知
　　（2）建筑构件图示认知
　　3.认知环境
　　（1）街道空间认知
　　（2）建筑肌理类型认知
　　（3）地形与植被认知
　　4.专业建筑表达
　　（1）建筑图纸表达
　　（2）建筑模型表达
　　（3）环境分析图表达

Subject Content
Cognition and Presentation
Training Objective
The course is the basic course of general professional knowledge for undergraduates of architecture. Task of the course is, on the one hand, allow students to cognize basic knowledge about physical building from an professional perspective, such as main building components and materials, basic constructional principles, spatial dimensions, and building environment etc.; and on the other hand, to better master such basic architectural knowledge through studying application of professional presentation method of architecture, and to lay down solid foundation for future in-depth study of professional knowledge.
Teaching Content
1. Cognizing building
(1) Surveying and drawing of partial elevation
(2) Surveying and drawing plans, profiles of building
(3) Surveying and drawing building components
2. Cognizing drawings
(1) Cognition to drawings of individual building
(2) Cognition to drawings of building components
3. Cognizing environment
(1) Cognition to street space
(2) Cognition to types of building texture
(3) Cognition to terrain and vegetation
4. Professional architectural presentation
(1) Presentation with architectural drawings
(2) Presentation with architectural models
(3) Presentation with environmental analysis charts

本科二年级

建筑设计（一）：休息与纪念空间设计

· 刘铨 冷天 王丹丹

课程类型：必修

学时学分：64学时 / 4学分

Undergraduate Program 2nd Year
ARCHITECTURAL DESIGN 1: REST AND MEMORIAL SPACE DESIGN · LIU Quan, LENG Tian, WANG Dandan
Type: Required Course
Study Period and Credits: 64 hours/4 credits

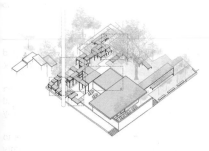

课题内容
　　校园休息廊与纪念亭设计
教学目标
　　训练基本建筑形式语言组织与身体感知（路径、视线、活动与尺度、光影、质感）的关系。通过建筑语言与感知形成一定的使用空间，并塑造出相应的空间氛围。
教学要点
　　1. 认知场地，学习提取场地条件：地形、朝向、交通、周边场地、建筑的功能与界面、地标、植被等。设计场地位于南京大学鼓楼校区健忠楼与旗杆广场之间的双杠运动区，用地范围详见地形图。
　　2. 形成网格，根据场地条件在网格基础上确定路径与功能区域。建筑应包含原有的布告栏与宣传栏功能，一位南京大学历史名人的纪念性空间，以及一个16人位咖啡店，包括不少于10人的室外或半室外座位。建筑面积不限，但场地内的绿地面积不应减少。建筑层数1层，可考虑使用屋顶平台。
　　3. 用基本围合限定路径与视线。
　　4. 根据视线、活动需要变化网格的尺度和空间的高度。
　　5. 组织光影，对空间限定做进一步细化。
　　6. 在基本结构构造需求下赋予空间材质，丰富场所感受。

Subject Content
Campus Veranda and Memorial Pavilion Design
Teaching Objective
Train the relationship between basic architectural form language organization and body sensing (path, sight, activity and dimension, light and shadow, texture). Form certain use space through architectural language and sensing, and create relevant space atmosphere.
Teaching Essential
1. Cognition site, study extracting site condition: landform, orientation, traffic, surrounding site, building function and interface, landmark, vegetation, etc.. The design site is located in the parallel bars sports area between Jianzhong Building and flagpole square in Gulou campus of Nanjing University, see the land scope in the topographic map.
2. Form grid, confirm path and functional area according to the site condition on the basis of the grid. The architecture shall include the original bulletin and billboard function, a memorial space for a historical celebrity of Nanjing University and a 16-seat café, including no less than 10 outdoor or semi-outdoor seats. The building area is not limited, however, the green area on the site shall not be reduced. It is 1 building floor, roof platform can be considered.
3. Use basic enclosure to limit path and sight.
4. Change grid dimension and space height according to sight, activity requirement.
5. Organize light and shadow, further refine the space limit.
6. Endow space material under the basic structure requirement, enrich place feeling.

本科二年级
建筑设计（二）：独立居住空间设计
· 冷天 刘铨 王丹丹
课程类型：必修
学时学分：64学时／4学分

Undergraduate Program 2nd Year
ARCHITECTURAL DESIGN 1: INDEPENDENT
LIVING SPACE DESIGN · LENG Tian,
LIU Quan, WANG Dandan
Type: Required Course
Study Period and Credits: 64 hours/4 credits

教学目标
综合运用建筑设计基础课程中的知识点，完成一个小型独立居住空间的设计方案。训练的重点一是内部空间的整合性设计，二是通过真实有效的建构设计将空间效果清晰地予以呈现。

设计要点
1. 场地与界面：场地从外部限定了建筑空间的生成条件。本次设计场地是传统老城内真实的建筑地块，面积在200m² 左右，单面或相邻两边临街，周边是1~2层的传统民居。主要要求学生从场地原有界面的角度来考虑设计建筑的形体、布局及其最终的空间视觉感受。
2. 功能与空间：本练习的功能设定为小型家庭独立居住空间，附设一个小型文化沙龙。
3. 流线与入口：在给定场地内生成建筑。一方面，内部空间的安排要考虑到与场地界面的关系，如街道界面的连续性、出入口的位置等。另一方面，空间的安排要考虑内部流线组织的合理性。
4. 尺度与感知：在空间形式处理中注意通过图示表达理解空间构成要素与人的空间体验的关系，主要包括尺度感和围合感。空间、楼梯、门窗等建筑构件的设计需满足基本建筑规范的要求。

Teaching Objectives
Comprehensively apply the knowledge in architectural design basic course, complete a small independent living space design plan. The training focus is integration design of internal space, and clearly showing space effect through real and effective tectonic design.
Key Point of Design
1. Site and interface: The site limits the generation condition of architectural space from outside. The design site is a real architectural plot in traditional old city, with the area of 200m², facing the street by one side or adjacent two sides, with traditional residence of 1~2 storeys surrounding. The students are mainly required to consider designing architectural form, layout and final space visual feeling from the perspective of original site interface.
2. Function and space: The function of the exercise is set as small family independent living space, attached with a small cultural saloon.
3. Streamline and entrance: Generate an architecture on the provided site. On one hand, internal space arrangement shall consider the relationship with the site interface, such as street interface continuity, position of access, etc.. On the other hand, space arrangement shall consider the rationality of internal streamline organization.
4. Dimension and sensing: The relationship between space composition and space experience shall be understood through graphic expression in space form treatment, mainly including sense of dimension and sense of enclosure. The design of space, stair, door, window, etc. shall meet the requirement of basic architectural code.

本科三年级
建筑设计（三）：小型公共建筑设计
· 周凌 童滋雨 窦平平
课程类型：必修
学时学分：72学时／4学分

Undergraduate Program 3rd Year
ARCHITECTURAL DESIGN 3: SMALL PUBLIC
BUILDING · ZHOU Ling, TONG Ziyu, DOU
Pingping
Type: Required Course
Study Period and Credits: 72 hours/4 credits

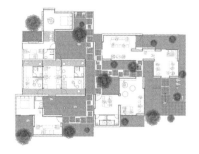

课题内容
乡村小型家庭旅馆扩建
教学目标
此课程训练解决两个基本问题：一是房屋结构、材料、构造等建造问题；二是基本起居、居住功能的平面功能排布问题。通过这一建筑设计课程的训练，使学生在学习设计的初始阶段就知道房子如何造起来，深入认识形成建筑的基本条件：结构、材料、构造原理及其应用方法，同时课程也面对地形、朝向、功能问题。训练核心是结构、材料、构造、基本功能，强化认识建筑结构、建筑构件、建筑围护等实体要素。
教学内容
观音殿村位于南京市江宁区秣陵街道，由于城乡统筹发展与乡村治理的需要，需要对村内现有房屋进行改造，部分房屋改造为乡村公共配套服务建筑，部分改造为对外服务与经营用房，部分改造为小型家庭旅馆。每个基地保留1~2栋老房子，改造为客房。另外在院子内进行加建设计，加建部分作为工坊、展示、客厅、餐厅等公共功能。
建筑层数1~2层，建筑限高：檐口高度不超过7.5m，总高不超过9m。平顶坡顶不限。要求充分考虑材料建造与实施的可能性。改造部分客房面积：单间20~30m²，套间30~45m²。加建公共部分面积约100~300m²。

Subject Content
Extension of a Country House as Holiday Inn
Training Objective
This course training solves two basic issues: one is the construction issue as housing structure, material, construction, etc.; another is the plan function arrangement issue as basic living, living function. Through the training of this architectural design course, students know how to build a house at initial stage of design study, deeply realize the basic condition for forming architecture: structure, material, construction principle and application method, at the same time, the course also aims at landform, orientation, function issue. The training core is structure, material, construction, basic function, reinforcing cognition of architectural structure, architectural component, architectural enclosure, etc..
Teaching Content
Guanyindian Village is located in Moling Street, Jiangning District, Nanjing. Due to the requirement of urban-rural integration development and countryside control, it is required to reform existing houses in the village, some houses are reformed into rural public supporting service building, some houses are reformed into external service and operation houses, some houses are reformed into small family inn. 1~2 old houses are reserved for each base, reformed into guestroom. In addition, additional construction is designed in the courtyard, the increased part is used as workshop, display, living room, dining room, etc..
The building has 1~2 floors, the building height is limited: the cornice height is not more than 7.5m, total height is not more than 9m. The flattop or slope crest is not limited. The possibility of material construction and implementation shall be fully considered. Reform part of guestroom area: single room 20~30m², suite 30~45m². The increased public area is about 100~300m².

本科三年级
建筑设计（四）：中型公共建筑设计
· 周凌 童滋雨 窦平平
课程类型：必修
学时学分：72学时／4学分

Undergraduate Program 3rd Year
ARCHITECTURAL DESIGN 4: PUBLIC BUILDING · ZHOU Ling, TONG Ziyu, DOU Pingping
Type: Required Course
Study Period and Credits: 72 hours/4 credits

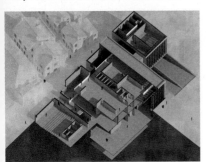

课题内容
利济巷纪念馆
教学目标
课程主题是"空间"和"氛围"。学习综合运用空间形式处理、流线秩序组织、光影明暗调和、材质组合搭配、公共景观配置和室内室外联系，共同营造空间氛围和气质。着重学习贯通空间的设计：内外、左右、前后、上下贯通。充分考虑身体在空间中的感受，以对模型的空间和光环境想象为手段反复推敲方案。不仅满足具体的使用需求，而且塑造精神性的场所——投射历史记忆，体现人性悲悯，向往和平宁静。同时，认识城市空间的复杂性，了解城市道路、文保建筑、地下建筑的规范要求。
教学内容
1.空间组织原则：空间组织要有明确特征，有明确意图，概念要清楚。并且满足功能合理、环境协调、流线便捷的要求。注意三种空间：聚散空间（门厅、出入口、走廊）；序列空间（单元空间）；贯通空间（平面和剖面上均需要贯通，内外贯通、左右前后贯通、上下贯通）。
2.空间类型：展览区：1 500 m²；报告厅：300 m²；研究空间：400 m²；行政办公空间：200 m²；服务空间：300 m²；其他空间：客梯、货用电梯各一部；向市民开放的室外公共空间，作为公共纪念和活动场地；地下停车场（停放50辆以下）。
建筑总面积3 000 m²，高度不高于10 m。

Subject Content
Lijixiang Memorial Hall
Training Objective
The course title is space and atmosphere. Learn comprehensive application of treating space form, organizing streamline order, tuning light and shadow, matching material combination, configuring public landscape and relating indoor and outdoor, and create space atmosphere and temperament. Focus on learning space-through design: internal and external, left and right, front and rear, top and bottom through. Fully consider the body feeling in space, repeatedly deliberate the plan by means of model space and light environment imagination. Not only meet specific service requirement, but also create spiritual place—project history memory, reflect pessimism, yearning for peace and tranquility. At the same time, realize complexity of urban space, know code requirement of urban road, heritage building, underground building.
Teaching Content
1. Space organization principles: Space organization requires distinctive characteristics, explicit intention, and clear concepts. It shall also meet the requirements of reasonable functions, coordinated environment, and convenient circulation. Attention shall be paid to three types of space: converging and diverging space (hallway, entrance and exit, corridor); sequence space (unit space); connecting space (connecting spaces are required on plans and profiles, internal-and-external connection, left-and-right, front-and-rear connections, up-and-down connection).
2. Space type: Exhibition area: 1 500 m²; report hall: 300 m²; research space: 400 m²; administrative office space: 200 m²; service space: 300 m²; other space: one passenger lift, one cargo lift; outdoor public open to the citizen, used as public memorial and activity place; underground parking lot (park less than 50 vehicle).
Total building area is 3 000 m², and the height is no higher than 10m.

本科三年级
建筑设计（五+六）：大型公共建筑设计
· 华晓宁 钟华颖 王铠
课程类型：必修
学时学分：144学时／8学分

Undergraduate Program 3rd Year
ARCHITECTURAL DESIGN 5&6: COMPLEX BUILDING · HUA Xiaoning, ZHONG Huaying, WANG Kai
Type: Required Course
Study Period and Credits: 144 hours/8 credits

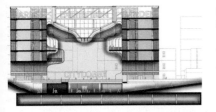

课题内容
城市建筑——社区商业中心＋活动中心设计
研究主题
实与空：关注城市中建筑实体与空间的相互定义、相互显现，将以往习惯上对于建筑本体的过度关注拓展到对于"之间"的空间的关注。
内与外：进一步突破"自身"与"他者"之间的界限，将个体建筑的空间与城市空间视为一个连续统，建筑空间即城市空间的延续，城市空间亦即建筑空间的拓展，两者时刻在对话、互动和融合。
层与流：不同类型的人和物的行为与流动是所有城市与建筑空间的基本框架，当代大都市中不同的流线在不同的高度上层叠交织，构成一个复杂的多维城市。必须首先关注行为和流线的组织，由此才发生出空间的系统和形态。
轴与界：城市纷繁复杂的形态表象之后隐含着秩序和控制性，并将成为新的形态介入。
教学内容
在用地上布置社区商业中心（约15 000 m²）、社区文体活动中心（约8 000 m²），并生成相应的城市外部公共空间。

Subject Content
Urban Buildings: Design of Community Business Center and Activity Center
Research Subject
Entity and space: Pay attention to mutual definition, mutual representation of architectural entity and space in cities, and extend traditional excessive attention to the building itself to the space "among them".
Interior and exterior: Further break through the boundary between "self" and "others", and consider space of individual buildings and urban space as a continuum.
Stack and flow: Behaviors and flows of different types of people and objects are the basic framework of all urban and building spaces, and a complex multi-dimensional city is formed by stacking up and interweaving of different flow lines at different altitudes in modern metropolis. We must pay attention to organization of behaviors and flow lines first, and then can generate system and morphology of space.
Axis and boundaries: Order and control are concealed behind the morphologic appearance of complexity of cities, which will be involved as new forms.
Teaching Content
Lay out community commercial center (about 15 000 m²) and community recreational and sports activities center (about 8 000 m²) on the land, and generate associated outdoor urban public spaces.

本科四年级
建筑设计（七）高层建筑设计
吉国华　华晓宁　尹航
课程类型：必修
学时学分：72 学时 / 4 学分

Undergraduate Program 4th Year
ARCHITECTURAL DESIGN 7: HIGH-RISING BUILDING · JI Guohua, HUA Xiaoning, YIN Hang
Type: Required Course
Study Period and Credits: 72 hours/4 credits

课题内容
高层办公楼设计

教学目标
高层办公建筑设计涉及城市、空间、形体、结构、设备、材料、消防等方面内容，是一项较复杂与综合的任务。本课题采取贴近真实实践的视角，教学重点与目标是帮助学生理解、消化以上涉及各方面知识，提高综合运用并创造性解决问题的技能。

教学内容
建筑容积率≤2.8，拟定建设指标为总建筑面积约3.8万 m²，地上建筑面积约2.8万 m²，地下建筑面积，约1.0万 m²。需规划合理流线，避免形成交通拥堵。
本项目是为江苏气象灾害监测预警及应急服务业务所需数据采集、传输、运算、分析、信息发布、灾害预警及防灾救灾服务等提供场所保障，需充分理解现代气象业务及科研用房的基本功能需求以及项目的特性，使功能布局科学合理。

课程特别要求
把建筑物理性能及绿色建筑设计的手段与方法应用于设计过程，在设计中予以表达。

Subject Content
Design of High-rise Office Building
Teaching Objective
High-rise office architectural design involves city, space, form, structure, equipment, material, fire protection, etc., as a complex and comprehensive task. This subject takes view close to realistic practice, and the teaching focus and objective are to help students understand, digest aforesaid knowledge, improve the skill of comprehensively applying and creatively solving problem.
Teaching Content
Building plot ratio ≤2.8, proposed construction index is total building area is about 38 000m², aboveground building area is about 28 000m², underground building area is about 10 000m². Rational streamline shall be planned to avoid traffic jam.
This project provides the place guarantee for data collection, transmission, operation, analysis, information release, disaster warning and disaster prevention and relief service required by meteorological disaster monitoring, warning and emergency service business in Jiangsu. Modern meteorological business and basic function requirement of scientific research housing and characteristics of the project shall be fully understood to make function layout scientific and rational.
Special requirement
Apply building physical performance and means and methods of green architectural design in design, and express in design.

本科四年级
建筑设计（八）城市设计
吉国华　胡友培　尹航
课程类型：必修
学时学分：72 学时 / 4 学分

Undergraduate Program 4th Year
ARCHITECTURAL DESIGN 8: URBAN DESIGN · JI Guohua, HU Youpei, YIN Hang
Type: Required Course
Study Period and Credits: 72 hours/4 credits

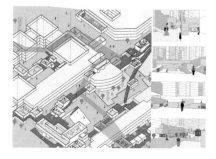

课题内容
南京碑亭巷地块旧城更新城市设计

教学目标
1.着重训练城市空间场所的创造能力，通过体验认知城市公共开放空间与城市日常生活场所的关联，运用景观环境的策略创造城市空间的特征。
2.熟练掌握城市设计的方法，培养从宏观整体层面处理不同尺度空间的能力，并有效地进行图纸表达。
3.理解城市更新的概念和价值；通过分析理解城市交通、城市设施在城市体系中的作用。
4.多人小组合伙，培养团队合作意识和分工协作的工作方式。

教学内容
1.设计地块位于南京市玄武区，总用地约为6.20hm²。地块内国民大会堂旧址、国立美术陈列馆旧址和北侧邮政厅大楼可保留，其余地块均需进行更新。地块周边有丰富的博物馆、民国建筑等文化资源，设计应对周边文化环境起到进一步提升作用。地块周边用地情况复杂，设计中需考虑与周边现状的相互影响。
2.本次设计的总容积率指标为2.0~2.5，建筑退让、日照等均按相关法规执行。
3.碑亭巷、石婆婆庵需保留，碑亭巷和太平北路之间现状道路可根据设计进行位置或线形调整。
4.地下空间除满足单一地块建筑配建的停车需求外，应综合考虑地上、地下城市一体化设计与综合开发。

Subject Content
Urban Design for Old Town Renovation of the Land Parcel of Beiting Lane in Nanjing
Training Objective
1. Emphasize the training on ability of creating urban spatial places, and create features of urban space with the strategy of landscape environment through experiencing and perceiving the links between urban public spaces and urban daily living places.
2. Master methodology of urban design, grasp the ability of handling spaces of different dimension at macro and integral level, and achieve effective representation with drawings.
3. Understand the concept and value of urban renovation; understand the role of urban traffic, urban facilities in the urban system through analysis.
4. Form partnership with several group members to cultivate awareness of teamwork and the working mode of collaboration.
Teaching Content
1. Land parcel of the design is located in Xuanwu District, Nanjing, covering an area of 6.20hm² approximately. Site of the former National Assembly Hall, site of the former National Art Gallery and the post office building at north side may be retained, other parts of the land parcel need to be renovated. There are abundant cultural resources such as museums and buildings constructed in the period of the Republic of China around the land parcel, so the design should further improve the cultural environment around it. Land use conditions around the land parcel is very complicated, so mutual influence with surrounding existing conditions must be taken into account in the design.
2. Gross plot ratio of the design is 2.0~2.5, and building setback and sunlight value shall comply with relevant laws and regulations.
3. Beiting Lane and Shipopo Nunnery are to be retained, and location and route of the existing road between Beiting Land and North Taiping Road may be adjusted according to the design.
4. For underground space, besides meeting the associated parking demand of buildings on the single land parcel, overall consideration shall be taken for integrated design and comprehensive development of urban spaces above and under the ground.

本科四年级
毕业设计
· 赵辰
课程类型：必修
学时学分：1 学期 /0.75 学分

Undergraduate Program 4th Year
THESIS PROJECT · ZHAO Chen
Type: Required course
Study Period and Credits: 1 term /0.75 credit

课题内容
历史街区复兴的整体设计，从室内到城市
教学内容
中国的历史城市与街区，作为物质性与非物质性文化遗产之依托，已经成为社会公众所认知的最有价值的城市空间。当下建筑师所充分需求参与的城市更新与再发展计划都将面临此类任务，也必将是合理应用所掌握的专业及多学科知识之挑战。
教学目标
掌握建筑设计基本的技能与知识（测绘、建模、调研、分析），并能对特定的地域和历史建筑进行深入的设计研究，根据社会发展的需求，提出改造和创造的可能。在选定的区域之历史与现状研究的基础上，选择相关重点节点，进行建筑的室内、单体、群体、外部空间、景观、交通等内容的整体设计。

Subject Content
Overall Design for Revival of Historical Block, from Interior to Urban
Teaching Content
Historical city and block in China, as the support of tangible and intangible cultural heritage, has become the most valuable urban space known to the social public. Current urban renewal and re-development plan that architects fully require participating in will face such task, and it will be the challenge of rationally applying professional and multi-disciplinary knowledge.
Training Objective
Master basic skill and knowledge of architectural design (surveying and mapping, modeling, investigating, analyzing), deeply design and research specific region and historical building, according to the requirement of social development, proper possibility of reform and creation. On the basis of researching history and current situation of selected region, select relevant node for overall design of interior, single, group, external space, landscape, traffic, etc..

本科四年级
毕业设计
· 钟华颖
课程类型：必修
学时学分：1 学期 /0.75 学分

Undergraduate Program 4th Year
THESIS PROJECT · ZHONG Huaying
Type: Required Course
Study Period and Credits: 1 term /0.75 credit

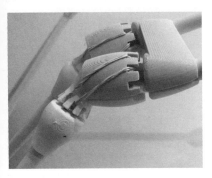

课题内容
基于可变密度构件的三维打印亭设计
教学内容
本课程为三维打印建筑的设计技术应用研究。三维打印技术正在向建筑尺度延伸，加工用于实际建造的建筑构件。探讨如何在设计的前期，针对三维打印的材料特性、加工方式，提出针对性的设计方法，加工满足结构受力与建筑效果的构件。课程内容以三维打印亭为题，利用数字设计找形技术寻求优化的建筑形态，利用几何细分生成建筑构件，再由可变密度三维打印技术对构件进行优化，获得适应整体建筑形态及受力要求的建筑构件。课程计划 15 周，分为基础技术学习、设计、加工建造、成果表达四个阶段。
教学目标
学习自由曲面建模、编程技术等参数化设计技术，完成课程作业要求的模型建模及程序编写。了解和掌握三维打印机等数控加工设备，完成一个三维打印实验构件。由整体形态细分确定单元构件形态尺寸及连接构造。生成适应受力要求的单元内部支撑结构，以不同打印密度平衡受力与透明度等设计要求。最终，每位组员完成一套设计成果及典型构件单元打印模型。优选一项设计共同完成 1/2 比例大小实物模型。

Subject Content
3D Printed Pavilion Design Based on Variable Density Component
Teaching Content
This course is design technology application and research of 3D printed architecture. 3D printing technology is being extended to architecture, processing building component in actual construction. Discuss how to propose targeted design method and process component that meets structure stress and building effect in early period of design according to the material characteristics and processing method of 3D printing. The course content is themed in 3D printed pavilion, utilizes digital design form finding technology to seek for optimized architectural form, utilizes geometric segment to generate building component, then optimizes the component through variable density 3D printing technology, so as to obtain the building component that adapts to the overall architectural form and stress requirement. The course lasts for 15 weeks, divided into basic technology, design, processing and construction, result expression.
Training Objective
Study Parametric design technology: free curve modeling, programming technology, etc.. Complete modeling and programming required by the course.Know and master CNC machining equipment as 3D printer, etc.Complete a 3D printed test component.Generate unit internal supporting structure that meets stress requirement, with different printing density balance stress and transparency requirements.Course result requirement: each team member shall complete a set of design result and typical component unit printing model. One design will be preferred to complete 1/2 physical model.

本科四年级
毕业设计
· 郜志
课程类型：必修
学时学分：1 学期 /0.75 学分

Undergraduate Program 4th Year
THESIS PROJECT · GAO Zhi
Type: Required course
Study Period and Credits: 1 term /0.75 credit

课题内容
绿色建筑技术系统评价的案例研究
教学内容
在绿色建筑设计过程中，建筑师一般根据以往经验整合各种绿建技术，预测其性能并评价其适用性。然而，由于气候及当地条件等差异导致相同的绿建技术对不同的建筑有不同的影响。多种绿建技术综合作用的适用性及潜在收益需要用系统的方法加以确定，用以指导实践。而且由于多种绿建技术一般是整合在一个或多个系统中，对单一技术的性能评价是相对困难的。故需要开发系统的方法和步骤评价绿建技术，包括绿建技术分类、开发绿色特征分析方法以及相关案例研究。
教学目标
本毕业论文题目首先对绿建技术进行系统分类，并通过分离变量法对绿色特征进行模块化，搭建基准平台，并以 EnergyPlus（或 Equest）为主要研究工具进行能耗模拟，最后对绿色特征进行影响分析和总结。

Subjective Content
Preliminary Study on Green Building Technology Systematic Evaluation Based on Case Studies
Teaching Content
In the design of green building, architect will integrate various green building technologies according to experience, forecast the performance and evaluate its applicability. However, same green technology has different influences on different architectures due to different climates and local conditions. The applicability and potential gain of comprehensive function of various green building technologies shall be confirmed in systematical method to guide the practice. Various green building technologies are generally integrated in one or several systems, and it is relatively difficult to evaluate the performance of single technology. Therefore, systematical method and step shall be developed to evaluate green building technology, including: classification of green building technology, green feature analytical method development and relevant case study.
Training Objective
The subject of the thesis classifies green building technology, modularizes green feature in variable separation method, builds up benchmarking platform, uses EnergyPlus (or Equest) as main research tool for energy consumption simulation, and finally analyzes and summarizes green feature.

本科四年级
毕业设计
· 吉国华
课程类型：必修
学时学分：1 学期 /0.75 学分

Undergraduate Program 4th Year
THESIS PROJECT · JI Guohua
Type: Required Course
Study Period and Credits: 1 term /0.75 credit

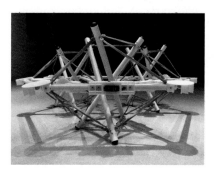

课题内容
基于标准单元的互动建筑界面："涌现"数字建筑设计
教学内容
在西方哲学史中，"涌现"思想的起源最早可以追溯到亚里士多德时期。"整体具有独特性，它们源自不同层级组织和整合之间连续的相互作用。""涌现"的系统特征包含并超越了组成这个系统的各个部分属性的集合。在自然界的形态中，"涌现"的系统结构随处可见，宏观的如风力形成的沙滩波浪形态和侵蚀的山谷形态，微观的如沸腾的水中冒出的大小不一的气泡和雪花或者冰凌结晶的形态。在计算机与人工智能领域，1980 年代末期出现诸如模拟鸟群飞行的 Boids 群聚模型的"涌现"行为模拟。进入 21 世纪，在环境设计领域，以英国建筑联盟学院、美国麻省理工学院等为首的高校开始借助计算机领域已经取得的"涌现"系统算法方面的成果，广泛地模拟自然界的"涌现"系统结构，在全球范围内掀起人居环境设计的泛参数化浪潮。
教学目标
本课题以基于标准单元的互动建筑界面设计为研究范围，通过实物模型制作来不断探索设计问题。通过分析学习当下的数字技术以及装配方法来将设计理念付诸建造实践。

Subject Content
Standard Unit Based Interactive Building Interface: "Surging" Digital Architectural Design
Teaching Content
In the history of western philosophy, the origin of "surging" can be traced back to the time of Aristotle. "The whole is special, from continuous mutual function between organization of different hierarchies and integration." The system characteristics of "surging" include and surpass the integration of attribute of all parts that compose this system. In the form of nature, "surging" system structure can be found everywhere, such as beach wave form formed by wind power and eroded valley form, bubbles in different sizes surged from boiling water or crystal form of snow flake or ice. In the field of computer and artificial intelligence, "surging" behavior simulation of Boids cluster model that simulates birds fly appeared in late 1980s. In the 21st century, in the field of environmental design, colleges headed by Architectural Association School of Architecture, Massachusetts Institute of Technology, etc. began to use the results in "surging" system algorithm in the field of computer to widely simulate the "surging" system structure of the nature, and created a worldwide pan-parameter tide of living environmental design.
Training Objective
The research scope of this subject is the interactive architectural interface design based standard unit, which is constantly explored through making material model. Design idea is applied in construction practice through analyzing and learning current digital technology and assembly method.

本科四年级
毕业设计
· 华晓宁
课程类型：必修
学时学分：1学期 /0.75 学分

Undergraduate Program 4th Year
THESIS PROJECT · HUA Xiaoning
Type: Required course
Study Period and Credits: 1 term /0.75 credit

课题内容
安徽省郎溪县定埠集镇更新
教学内容
　　新型城镇化进程中的乡镇更新与复兴是当前我国建筑学所面临的主要课题之一。乡镇复兴并非仅仅意味着物质空间环境的重建或改造，更重要的是必须赋予乡镇生活以新的活力。通过事件及其所驱动的城镇景观演变，推动城镇的更新与复兴，是一种重要的空间活化策略。
教学目标
　　掌握建筑设计基本的技能与知识（测绘、建模、调研、分析），并能对特定的地域和历史建筑进行深入的设计研究，根据社会发展的需求，提出改造和创造的可能。在选定的区域之历史与现状研究的基础上，选择相关重点节点，进行建筑的室内、单体、群体、外部空间、景观、交通等内容的整体设计。

Subjective Content
Update of Dingbu Town, Liangxi County, Anhui Province
Teaching Content
Township upgrading and revival in new type urbanization process is one of the main subjects for current Chinese architecture. Township revival does not only mean reconstruction or reform of physical space and environment, but also new vigor endowed to township life. Promoting township upgrading and revival through events and urban landscape evolution it drives is an important space activating policy.
Training Objective
Master basic skill and knowledge of architectural design (surveying and mapping, modeling, investigating, analyzing), deep design and research specific region and historical building, according to the requirement of social development, proper possibility of reform and creation. On the basis of researching history and current situation of selected region, select relevant node for overall design of interior, single, group, external space, landscape, traffic, etc.

本科四年级
毕业设计
· 胡友培
课程类型：必修
学时学分：1学期 /0.75 学分

Undergraduate Program 4th Year
THESIS PROJECT · HU Youpei
Type: Required Course
Study Period and Credits: 1 term /0.75 credit

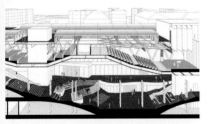

课题内容
滨江绿道工业遗存活化与文化综合体设计
教学内容
　　在历史感稀薄的当代中国城市中，存在着各种零星的历史片段。它们呈现为某种物质形态或集体记忆，与城市日常公共生活构成层叠、并置的松散关系。历史与当下、宏大与日常、官方与市民，这是中国城市在历史文化层面的二元命题。南京是一座临江城市，长江曾一度淡出日常城市生活。近年来，随着滨江绿道的建设，这一情况有所改善，江堤岸线重新回归城市。绿道中的一段，一座尺度宏阔的社会主义工业遗存与一处红色集体记忆场所不期而遇。异质的城市空间、戏剧化的浮现、现实充满穿越与违和，却又让人心生期待。这样一个异质的场地，对于城市而言意味着什么？锈迹斑斑的旧厂房与略显陈旧的展陈，可能蕴涵什么样的潜力？还能够以何种方式，延续历史叙事的同时，贡献新的城市价值？——这是一个城市建筑学的问题。
教学目标
　　本次毕业设计题目，具有城市建筑的多种复杂要素，涉及历史遗存、城市记忆、公共景观、单体改造等诸多层次，需要学习并综合运用多方面专业知识，具有一定的综合性和设计难度。同时，作为真实项目，项目将面对来自业主诉求、规划控规要点、法律法规等现实层面的需求与挑战。本毕设将以贴近实践的视角，培养在真实语境中进行建筑创作的综合能力。

Subjective Content
Design of Industrial Relics Activation and Cultural Complex of Riverside Greenway
Teaching Content
In contemporary Chinese cities with thin sense of history, there are various sporadic history fragments. They are reflected in some physical form or collective memory, and constitute overlapped, juxtaposed loose relationship with urban daily public life. History and presence, magnificence and everydayness, government and citizen, these is the binary theme of Chinese city on the level of history and culture. Nanjing is a riverside city, and Yangtze River once faded out daily urban life. In recent years, with the construction of riverside Greenway, this situation is improved, riverbank line returns to the city. A section of the Greenway, a vast socialist industrial relic and a red collective memory meet by chance. Heterogeneous urban space, dramatic emergence, reality is full of cross and inharmony, but has something to look forward to.What such a heterogeneous place means to the city? Rusty old factory building and old display, what potential they may contain? How new urban value can be contributed while continuing the historical narration? This is a question about urban architecture.
Training Objective
The graduation project subject has various complex elements of urban architecture, involves historic relic, urban memory, public landscape, unit reform, etc., requires study and comprehensive application of various professional knowledge, and has certain comprehensiveness and design difficulty. At the same time, as a real project, the project will face the requirement and challenge from owner's appeal, planning control essentials, laws, regulations, etc.. The graduation design will cultivate comprehensive ability of architectural creation in real context with the perspective close to the practice.

本科四年级
毕业设计
· 周凌
课程类型：必修
学时学分：1 学期 /0.75 学分

Undergraduate Program 4th Year
THESIS PROJECT · ZHOU Ling
Type: Required Course
Study Period and Credits: 1 term /0.75 credit

课题内容
江苏特色田园乡村规划与建筑设计
教学内容
乡村不仅是传统的农业生产地和农民聚集地，还兼具经济、社会、文化、生态等多重价值和功能。经过多年不懈努力，江苏乡村建设发展不断迈上新台阶，但从总体上看，乡村仍然是高水平全面小康的突出短板。特别是在新型城镇化快速发展进程中，乡村面临着资源外流、活力不足、公共服务短缺、人口老化和空心化、乡土特色受到冲击破坏等问题和挑战，迫切需要重塑城乡关系，遵循发展规律，坚持走符合乡村实际的路子，努力建设立足乡土社会、富有地域特色、承载田园乡愁、体现现代文明的特色田园乡村，加快实现乡村的发展与复兴，推动农业现代化与城乡发展一体化互促共进。特色田园乡村建设作为推进农业供给侧结构性改革、在全国率先实现农业现代化的新路径，作为传承乡村文化、留住乡愁记忆的新载体，具有重要意义。
教学目标
本次毕业设计选择南京江宁徐家院、南京汤山、淮安金湖等特色田园乡村项目，进行实际参与，完成带有研究性和前瞻性的人居环境规划设计方案。

Subject Content
Jiangsu Characteristic Countryside Planning and Architectural Design
Teaching Content
The countryside is not only the traditional agricultural production place and farmer gathering place, but also has multiple value and functions in economy, society, culture, ecology, etc.. Over years of unremitting efforts, the rural construction development in Jiangsu has made progress, however, on the whole, countryside is still the obvious short slab for high-level all-round well-being, especially in the rapid development of new urbanization. Countryside faces the problem and challenge as resource outflow, insufficient vigor, short public service, population aging and hollowing, impact and damage of countryside characteristics, etc., it is in sore need of rebuilding urban-rural relationship, abiding by development rule, insisting on taking the road that meets the rural actuality, making efforts to construct characteristic countryside that bases on countryside society, is rich in regional characteristics, carries countryside nostalgia, reflects modern civilization, accelerating the realization of rural development and revival, promoting integrated and mutual promotion of agricultural modernization and urban-rural development. Construction of characteristic countryside, as the new path to promote agricultural supply side structural reform, takes the lead in realizing agricultural modernization, as the new carrier of inheriting rural culture, reserving nostalgia memory, has important meaning.
Training Objective
The graduation project selects characteristic countryside project as Xujiayuan in Jiangning District, Nanjing, Tangshan in Nanjing, Jinhu in Huaian, etc. for actual participation, and completes research-based prospective living environment planning design scheme.

本科四年级
毕业设计
· 丁沃沃
课程类型：必修
学时学分：1 学期 /0.75 学分

Undergraduate Program 4th Year
THESIS PROJECT · DING Wowo
Type: Required Course
Study Period and Credits: 1 term /0.75 credit

设计题目
福建长汀历史文化名城：城市更新与建筑设计
教学内容
长汀县位于福建省西南部闽赣边境，依卧龙山而傍汀江，城内保存了众多的寺庙、祠堂和传统大宅院，既有众多的传统木构建筑和夯土建筑，又有民国时期的闽南洋房，1994年被评为国家历史文化名城。历史上，长汀县被称为客家首府，是客家文化重要的聚集地。同时，长汀又是重要的红色根据地，共和国建国时期的主要的领导人都曾在此地逗留或小住，古城内保留着多处红色文化遗址。古城周围群山环绕，所以又被评为国家级生态名城。虽然古城拥有独特的旅游资源，但是近年来经济发展的诉求使古城面临着巨大的压力，保护古城和古建筑早已不只是技术的问题，单纯的保护早已使城市不堪重负，以保护为目的的城市设计被认作在保护的基础上给城市带来活力的有效途径，而建筑设计是完成目标的最终手段。
教学目标
本课题以长汀古城历史街区为建筑设计研究范围，通过调研和访谈理解设计问题，通过测绘和分析学习传统建筑的类型和优势以及地方建造的工法。阅读文献资料相关理论，并通过具体的设计研究与实验将所学转化为学理层面的知识和设计方法。

Subject Content
Famous Historic and Cultural City in Changting County of Fujian Province: Urban Renewal and Architectural Design
Teaching Content
Changting County is located at the frontier of Fujian and Jiangxi in the southwest of Fujian Province, backing on Wolongshan and close to Tingjiang, a lot of temples, ances tral halls and traditional courtyards are reserved in the city, including both traditional wooden architectures and cobs, and also western style houses during the Republic of China, and Changting County was assessed as famous historic and cultural city in 1994. In history, Changting County is known as the capital of the Hakkas, as the important gathering place of Hakka culture. At the same time, Changting is also an important red base, where main leaders during the founding of China stayed or temporarily lived, and a lot of red cultural sites are reserved in the ancient city. The ancient city is surrounded by mountains, and was assessed as famous ecological city. Although the ancient city has unique tourism resource, the appeal of economic development in recent years has made the city face huge pressure, protecting ancient city and ancient architecture is not only a technical issue, pure protection has overwhelmed the city, protection-purpose urban design is believed the effective approach to bring vigor to the city on the basis of protection, and architectural design is the final means to realize this objective.
Training Objective
This subject takes the historical block of Changting ancient city as the research scope of architectural design, and understands design issue through survey and interview. Type and advantage of traditional architecture and local construction method are learned through surveying, mapping and analyzing. Read literatures and relevant theories, and transform what is learned into knowledge and design method on scientific level through specific design research and experiment.

研究生一年级
建筑设计研究（一）：基本设计
· 傅筱
课程类型：必修
学时学分：40学时／2学分

Graduate Program 1st Year
DESIGN STUDIO 1: BASIC DESIGN · FU Xiao
Type: Required Course
Study Period and Credits: 40 hours/2 credits

课题内容
　　宅基地住宅设计
教学目标
　　课程从"场地、空间、功能、经济性"等建筑的基本问题出发，通过宅基地住宅设计，训练学生对建筑逻辑性的认知，并让学生理解有品质的设计是以基本问题为基础的。
研究主题
　　设计的逻辑思维
设计内容
　　在A、B两块宅基地内任选一块进行住宅设计。

Subject Content
Homestead Housing Design
Training Objective
The course starts from fundamental issues of architecture such as "site, space, function, and economy", aims to train students to cognize architectural logics, and allow them to understand that quality design is based on such fundamental issues.
Research Subject
Logical Thinking of Design
Teaching Content
Select one from two homesteads A and B, conduct housing design.

研究生一年级
建筑设计研究（一）：基本设计
· 张雷
课程类型：必修
学时学分：40学时／2学分

Graduate Program 1st Year
DESIGN STUDIO 1: BASIC DESIGN · ZHANG Lei
Type: Required Course
Study Period and Credits: 40 hours/2 credits

课题内容
　　传统乡村聚落复兴研究
教学目标
　　课程从"环境""空间""场所"与"建造"等基本的建筑问题出发，对乡村聚落肌理、建筑类型及其生活方式进行分析研究，通过功能置换后的空间再利用，从建筑与基地、空间与活动、材料与实施等关系入手，强化设计问题的分析，强调准确的专业性表达。通过设计训练，达到对地域文化以及建筑设计过程与方法的基本认识与理解。
研究主题
　　乡土聚落／民居类型／空间再利用／建筑更新／建造逻辑
教学内容
　　对选定的乡村聚落进行调研，研究功能置换和整修改造的方法和策略，促进乡村传统村落的复兴。

Subject Content
Research on Revitalization of Traditional Rural Settlements
Training Objective
This course starts with basic architectural problems like "environment" "space" "place" and "construction", analyzes and studies the texture, architectural type and life style of rural settlement, strengthens the analysis of design problems and emphasizes accurate professional expression from the relationship between building and base, space and activities, and material and implementation through spatial reuse after function replacement, to obtain basic knowledge and understanding of the regional culture as well as the process and methods of architectural design.
Research Subject
Rural settlement / types of folk house / reutilization of space / building renovation / constructional logic
Teaching Content
Conduct investigation and research on selected rural settlement, study methodology and strategy of function replacement and renovation and improvement, and promote revitalization of traditional rural villages.

研究生一年级
DESIGN STUDIO 1: CONCEPTUAL DESIGN · LU Andong
· 鲁安东
课程类型：必修
学时学分：40 学时／2 学分

Graduate Program 1st Year
DESIGN STUDIO 1: CONCEPTUAL DESIGN · LU Andong
Type: Required Course
Study Period and Credits: 40 hours/2 credits

课题内容
记忆 · 叙事 · 基于场所的设计
研究主题
在我们这个时代，建筑学不应再是铺陈记忆的艺术，而是一种提醒的艺术。它为记忆的显影提供参照系，帮助人与场所之间建立起深沉的情感归属，进而得以触及人的灵魂。它是一种叙事的艺术。
记忆既是对当代建筑学的巨大挑战，也是其物质建造的终极命题。我们需要新的形式，叙事的形式，来涉足未知的记忆领域。
从记忆的角度来说，物质、活动、感动与存在之间不存在清晰的边界划分。它们共同构成了一次浮现，通过现在的再想象连接起过去与未来。
记忆的建筑学关注记忆在场所中的发生机制，而不是对记忆内容的选择性再现。记忆的建筑学关注主体参与的精神体验，而不是映射着环境的感知体验。它需要超越了物质、空间与符号的新建筑语言，来沉浸、互动、质询与映射主体。正如现代主义背离了语言，用感知替代了人与场所之间的意义关联，记忆的建筑学将重新回到语言的领域，并召唤叙事的心灵。
场所记忆的意义建构应该是开放的，而不是被给予的。设计和技术，应该帮助人找回自我与世界的归属感。它们必须走向存在的维度。这是当代建筑学的使命。
教学内容
此次"概念设计"课程选择了南京长江大桥作为研究对象。对南京长江大桥的记忆交织着宏大的历史记忆与亲密的个体记忆，曾是许许多多人建设、学习、旅行、工作、生活乃至想象的一部分。本课程从大桥记忆入手，探讨空间记忆的特征、发生机制和意义建构，最终设计一个当代的、公共的和富有创造性的记忆场所。

Subject Content
Memory, Narrative and Place Memory
Research Subject
At this time, architecture should not be an art to display memory, but an art of reminding. It provides reference to manifest memory, helps people and place establish profound emotional belonging, so as to touch the soul. It is an art of narration.
Memory is a huge challenge to contemporary architecture, and also the ultimate proposition of its material building. We need new form, form of narration to step in the unknown field of memory.
From the perspective of memory, substance, activity, touch and existence have no clear boundary. They compose an appearance, and connect the past and the future through current re-imagination.
Architecture of memory concerns the occurrence mechanism of memory in the place instead of selective reproduction of memory content. Architecture of memory concerns the spiritual experience of entity participation instead of reflecting the emotional experience of environment. It needs new architectural language that surpasses substance, space and symbol to immerse, interact, question and map the entity. Just as modernism deviated from language, perception replaces the meaning relation between people and place, architecture of memory will return to the field of language and recall the soul of narration.
The meaning construction of place memory should be open instead of being given. Design and technology should help people retrieve the sense of belonging of self and world. They must walk to the dimension of existence. This is the mission of contemporary architecture.
Teaching Content
In the course of conceptual Design, Nanjing Yangtze River Bridge is selected as the research object. The memory of Nanjing Yangtze River Bridge interweaves grand history memory and close personal memory, and it is a part of construction, learning, travel, work, life and even imagination of many people. This course starts with the bridge memory, discusses on the characteristics, occurrence mechanism and meaning construction of space memory, and finally designs a contemporary, public and creative memory place.

研究生一年级
建筑设计研究（二）：建构设计
· 傅筱
课程类型：必修
学时学分：40 学时／2 学分

Graduate Program 1st Year
DESIGN STUDIO 2: CONSTRUCTIONAL DESIGN · FU Xiao
Type: Required Course
Study Period and Credits: 40 hours/2 credits

课题内容
"基础设计"的深化与发展：以基础设计案例为基础进行深化设计，要求达到节点大样表达深度。
教学目标
1. 训练学生对设计概念与结构、构造设计关联性的认知。
2. 训练学生对一个"完整空间形态"建造的认知。
教学内容
第一阶段，学习以往案例以及 Karama 受力分析软件。
第二阶段，进行结构分析研究，理解结构体系与空间意图表达的关系。
第三阶段，进行构造设计研究，理解构造与设计意图表达的关系。
第四阶段，成果制作。

Subject Content
Deepening and development of Basic Design: deep design on the basis of basic design case, meet node detail expression depth.
Research Objective
1. Train students to realize relevance of design concept and structure, construction design.
2. Train students to realize a "complete space form" construction.
Teaching Content
At the first stage, study past cases and Karama stress analysis software.
At the second stage, analyze and research structure, understand relationship between structure system and space expression.
At the third stage, design and research construction, understand the relationship between construction and design intention.
At the fourth stage, make result.

研究生一年级
建筑设计研究（二）：城市设计
· 胡友培
课程类型：必修
学时学分：40 学时／2 学分

Graduate Program 1st Year
DESIGN STUDIO 2: URBAN DESIGN
DESIGN · HU Youpei
Type: Required Course
Study Period and Credits: 40 hours/2 credits

课题内容
　　城市更新与设计
研究目的
　　创造一种线性的，具有城市尺度的物质系统，可以是建筑、基础设施、景观以及任意的混杂体。由于其超长的尺度，相应地需具备城市全局层面的再生与更新意义。
　　创造因地制宜的局地城市小生态。随着线性系统的展开，不断地与局地的城市片段建立关联与施加影响，实现其在局地层面的更新意义。
　　最后，创造一种新的城市物质形象以及形象背后的一种可能的城市建筑与城市生活。
教学内容
　　宁芜线南京段改线方案已获国家铁路总局批准，改线搬迁在即。新线路将在古雄站折向东北，经南京高铁南站，沿绕城高速外侧绕行至沧波门，避开主要城区。铁路外迁后将为城市遗留下一条极长又极细的城市空地。
　　对于这块特殊用地的更新与利用，城市规划部门提出的意向方案是沿原路段建设地铁8号线。由于线路穿越密集城区，大部分路段为下埋形式。其地表的使用仍然悬而未决。
　　通过一种建筑学的设计思维与空间形式，发掘该用地的潜力，同时赋予其建筑学的品质与城市的意义，正是本次课程着力探讨的议题。

Subject Content
Urban Renewal and Design
Teaching Objectives
Create a linear material system with urban dimension, it can be architecture, infrastructure, landscape and any mixed body. The ultra-long dimension shall have regeneration and renewal meaning on the overall urban level.
Create local urban micro-ecology that meets local condition. With the expansion of linear system, constantly relate and affect local urban section, realize the renewal meaning on local level.
Finally, create a new urban material image and a possible urban architecture and urban life behind the image.
Teaching content
Nanjing-Wuhu Line Nanjing Section reform plan has been approved by National Railway Administration. New line will turn to northeast at Guxiong Station, via Nanjing High-speed Railway South Station, detour to Cangbomen along outside the belt expressway, so as to avoid main urban area. After the railway is moved out, an extremely long and thin urban clearing will be left.
For the renewal and utilization of this special land, the intention plan proposed by urban planning department is to construct subway No.8 line along original road section. The line runs through dense urban area, most road sections are buried. The use of earth's surface is still pending.
through an architectural design thinking and space form, dig the potential of this land, endow quality of architecture and meaning of city, which is just the topic of this course.

研究生一年级
建筑设计研究（二）：城市设计
· 鲁安东
课程类型：必修
学时学分：40 学时／2 学分

Graduate Program 1st Year
DESIGN STUDIO 2 : URBAN
DESIGN · LU Andong
Type: Required Course
Study Period and Credits: 40 hours/2 credits

课题内容
　　面向增强场所的水乡聚落更新设计研究
研究主题
　　本课程以"场所"为切入点，在人类学视野下研究江南地区的水乡聚落，进而探索综合当代媒介与设计来创造增强场所的方法，以及这些方法的可能运用。
研究计划
　　学生将分为兴化市沙沟镇和江阴市长泾镇两个田野工作坊，分别研究"社会性增强场所"和"文化性增强场所"这两个问题。

Subject Content
Research on Renewal Design of Waterside Settlement for Enhanced Sites
Research Subject
This course takes "site" as the entry point, studies the waterside settlements in the south of the lower reaches of the Yangtze River from the perspective of anthropology, and then explores the methods to create enhanced sites by integrating contemporary media and design, as well as the possible application of these methods.
Research Plan
Students will be assigned to two field workshops in Shagou Town of Xinghua City and Changjing Town of Jiangyin City to study "socially enhanced sites" and "culturally enhanced sites" respectively.

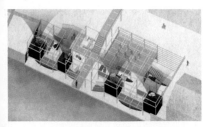

建筑理论课程
ARCHITECTURAL THEORY COURSES

本科二年级
建筑导论·赵辰 等
课程类型：必修
学时/学分：36学时/2学分

Undergraduate Program 2nd Year
INTRODUCTION TO ARCHITECTURE • ZHAO Chen, etc.
Type: Required Course
Study Period and Credits: 36 hours / 2 credits

课程内容
1. 建筑学的基本定义
第一讲：建筑与设计/赵辰
第二讲：建筑与城市/丁沃沃
第三讲：建筑与生活/张雷
2. 建筑的基本构成
（1）建筑的物质构成
第四讲：建筑的物质环境/赵辰
第五讲：建筑与节能技术/秦孟昊
第六讲：建筑与生态环境/吴蔚
第七讲：建筑的材料观/王丹丹
（2）建筑的文化构成
第八讲：建筑与人文、艺术、审美/赵辰
第九讲：建筑与环境景观/华晓宁
第十讲：建筑的环境智慧/窦平平
第十一讲：建筑与身体经验/鲁安东
（3）建筑师职业与建筑学术
第十二讲：建筑与表现/赵辰
第十三讲：建筑与几何形态/周凌
第十四讲：建筑与数字技术/钟华颖
第十五讲：建筑师的职业技能与社会责任/傅筱

Course Content
I Preliminary of architecture
1. Architecture and design / ZHAO Chen
2. Architecture and urbanization / DING Wowo
3. Architecture and life / ZHANG Lei
II Basic attribute of architecture
II-1 Physical attribute
4. Physical environment of architecture / ZHAO Chen
5. Architecture and energy saving / QING Menghao
6. Architecture and ecological environment / WU Wei
7. Architecture with Materialization / Wang Dandan
II-2 Cultural attribute
8. Architecture and civilization, arts, aesthetic / ZHAO Chen
9. Architecture and landscaping environment / HUA Xiaoning
10. Environmental Intelligence in Architecture / Dou Pingping
11. Architecture and body / LU Andong
II-3 Architect: profession and academy
12. Architecture and presentation / ZHAO Chen
13. Architecture and geometrical form / ZHOU Ling
14. Architectural and digital technology / ZHONG Huaying
15. Architect's professional technique and responsibility / FU Xiao

---

本科三年级
建筑设计基础原理·周凌
课程类型：必修
学时/学分：36学时/2学分

Undergraduate Program 3rd Year
BASIC THEORY OF ARCHITECTURAL DESIGN
• ZHOU Ling
Type: Required Course
Study Period and Credits: 36 hours / 2 credits

教学目标
本课程是建筑学专业本科生的专业基础理论课程。本课程的任务主要是介绍建筑设计中形式与类型的基本原理。形式原理包含历史上各个时期的设计原则，类型原理讨论不同类型建筑的设计原理。
课程要求
1. 讲授大纲的重点内容；
2. 通过分析实例启迪学生的思维，加深学生对有关理论及其应用、工程实例等内容的理解；
3. 通过对实例的讨论，引导学生运用所学的专业理论知识，分析、解决实际问题。
课程内容
1. 形式与类型概述
2. 古典建筑形式语言
3. 现代建筑形式语言
4. 当代建筑形式语言
5. 类型设计
6. 材料与建造
7. 技术与规范
8. 课程总结

Training Objective
This course is a basic theory course for the undergraduate students of architecture. The main purpose of this course is to introduce the basic principles of the form and type in architectural design. Form theory contains design principles in various periods of history; type theory discusses the design principles of different types of building.
Course Requirement
1. Teach the key elements of the outline;
2. Enlighten students' thinking and enhance students' understanding of the theories, its applications and project examples through analyzing examples;
3. Guide students using the professional knowledge to analysis and solve practical problems through the discussion of examples.
Course Content
1. Overview of forms and types
2. Classical architecture form language
3. Modern architecture form language
4. Contemporary architecture form language
5. Type design
6. Materials and construction
7. Technology and specification
8. Course summary

本科三年级
居住建筑设计与居住区规划原理·冷天 刘铨
课程类型：必修
学时/学分：36学时/2学分

Undergraduate Program 3rd Year
THEORY OF HOUSING DESIGN AND RESIDENTIAL PLANNING • LENG Tian, LIU Quan
Type: Required Course
Study Period and Credits: 36 hours / 2 credits

课程内容
第一讲：课程概述
第二讲：居住建筑的演变
第三讲：套型空间的设计
第四讲：套型空间的组合与单体设计（一）
第五讲：套型空间的组合与单体设计（二）
第六讲：居住建筑的结构、设备与施工
第七讲：专题讲座：住宅的适应性，支撑体住宅
第八讲：城市规划理论概述
第九讲：现代居住区规划的发展历程
第十讲：居住区的空间组织
第十一讲：居住区的道路交通系统规划与设计
第十二讲：居住区的绿地景观系统规划与设计
第十三讲：居住区公共设施规划、竖向设计与管线综合
第十四讲：专题讲座：住宅产品开发
第十五讲：专题讲座：住宅产品设计实践
第十六讲：课程总结、考试答疑

Course Content
Lect. 1: Introduction of the course
Lect. 2: Development of residential building
Lect. 3: Design of dwelling space
Lect. 4: Dwelling space arrangement and residential building design (1)
Lect. 5: Dwelling space arrangement and residential building design (2)
Lect. 6: Structure, detail, facility and construction of residential buildings
Lect. 7: Adapt ability of residential building, supporting house
Lect. 8: Introduction of the theories of urban planning
Lect. 9: History of modern residential planning
Lect. 10: Organization of residential space
Lect. 11: Traffic system planning and design of residential area
Lect. 12: Landscape planning and design of residential area
Lect. 13: Public facilities and infrastructure system
Lect. 14: Real estate development
Lect. 15: The practice of residential planning and housing design
Lect. 16: Summary, question of the test

研究生一年级
现代建筑设计基础理论 · 张雷
课程类型：必修
学时/学分：18学时/1学分

Graduate Program 1st Year
PRELIMINARIES IN MODERN ARCHITECTURAL DESIGN · ZHANG Lei
Type: Required Course
Study Period and Credits: 18 hours/1 credit

教学目标
建筑可以被抽象到最基本的空间围合状态来面对它所必须解决的基本的适用问题，用最合理、最直接的空间组织和建造方式去解决问题，以普通材料和通用方法去回应复杂的使用要求，是建筑设计所应该关注的基本原则。

课程要求
1. 讲授大纲的重点内容；
2. 通过分析实例启迪学生的思维，加深学生对有关理论及其应用、工程实例等内容的理解；
3. 通过对实例的讨论，引导学生运用所学的专业理论知识，分析、解决实际问题。

课程内容
1. 基本建筑的思想
2. 基本空间的组织
3. 建筑类型的抽象与还原
4. 材料的运用与建造问题
5. 场所的形成及其意义
6. 建筑构思与设计概念

Training Objective
The architecture can be abstracted into spatial enclosure state to encounter basic application problems which must be settled. Solving problems with most reasonable and direct spatial organization and construction mode, and responding to operating requirements with common materials and general methods are basic principle concerned by building design.

Course Requirement
1. To teach key contents of syllabus;
2. To inspire students' thinking, deepen students' understanding on such contents as relevant theories and their application and engineering example through case analysis.
3. To guide students to use professional theories to analyze and solve practical problems through discussion of instances.

Course Content
1. Basic architectural thought
2. Basic spacial organization
3. Abstraction and restoration of architectural types
4. Utilization and construction of material
5. Formation of site and its meaning.
6. Architectural conception and design concept

研究生一年级
研究方法与写作规范 · 鲁安东 胡恒 郜志
课程类型：必修
学时/学分：18学时/1学分

Graduate Program 1st Year
RESEARCH METHOD AND THESIS WRITING · LU Andong HU Heng GAO Zhi
Type: Required Course
Study Period and Credits: 18 hours/1 credit

教学目标
面向学术型硕士研究生的必修课程。它将向学生全面地介绍学术研究的特性、思维方式、常见方法以及开展学术研究必要的工作方式和写作规范。考虑到不同领域研究方法的差异，本课程的授课和作业将以专题的形式进行组织，包括建筑研究概论、设计研究、科学研究、历史理论研究4个模块。学生通过各模块的学习可以较为全面地了解建筑学科内主要的研究领域及相应的思维方式和研究方法。

课程要求
将介绍建筑学科的主要研究领域和当代研究前沿，介绍"研究"的特性、思维方式、主要任务、研究的工作架构以及什么是好的研究，帮助学生建立对"研究"的基本认识；介绍文献检索和文献综述的规范和方法；介绍常见的定量研究、定性研究和设计研究的工作方法以及相应的写作规范。

课程内容
1. 综述
2. 文献
3. 科学研究及其方法
4. 科学研究及其写作规范
5. 历史理论研究及其方法
6. 历史理论研究及其写作规范
7. 设计研究及其方法
8. 城市规划理论概述

Training Objective
It is a compulsory course to MA. It comprehensively introduces features, ways of thinking and common methods of academic research, and necessary manner of working and writing standard for launching academic research to students. Considering differences of research methods among different fields, teaching and assignment of the course will be organized in the form of special topic, including four parts: introduction to architectural study, design study, scientific study and historical theory study. Through the study of all parts, students can comprehensively understand main research fields and corresponding ways of thinking and research methods of architecture.

Course Requirement
The course introduces main research fields and contemporary research frontier of architecture, features, ways of thinking and main tasks of "research", working structure of research, and definition of good research to help students form basic understanding of "research". The course also introduces standards and methods of literature retrieval and review, and working methods of common quantitative research, qualitative research and design research, and their corresponding writing standards.

Course Content
1. Review
2. Literature
3. Scientific research and methods
4. Scientific research and writing standards
5. Historical theory study and methods
6. Historical theory study and writing standards
7. Design research and methods
8. Overview of urban planning theory

城市理论课程
URBAN THEORY COURSES

Undergraduate Program 4th Year
THEORY OF URBAN DESIGN • HU Youpei
Type: Required Course
Study Period and Credits: 36 hours / 2 credits

Course Content
Lect. 1. Introduction
Lect. 2. Technical terms: terms of urban planning, urban morphology, urban traffic and fire protection
Lect. 3. Urban design methods — documents analysis: urban planning and policies; relative documents; document analysis techniques and skills
Lect. 4. Urban design methods — data analysis: data analysis of demography, traffic flow, function distribution, visual and building height; modelling urban spatial data
Lect. 5. Urban design methods — classification of urban fabrics: introduction of urban fabrics; urban fabrics and floor area ratio; urban fabrics and open space; urban fabrics and traffic flow; criteria system of urban green space
Lect. 6. Urban design methods — organization of urban road network: introduction; urban road network and urban function; urban road network and urban space; urban road network and civic facilities; design of urban road section
Lect. 7. Urban design methods — representation skills of urban Design: mapping and analysis; conceptual diagram; analytical representation of urban design; representation of detail design; spatial representation of urban design
Lect. 8. Brief history and theories of urban design: historical meaning of urban design; connotation of urban design theories
Lect. 9. Form of urban road network: typology, structure and evolution of road network; road network and urban fabrics
Lect. 10. Urban space: typology, structure, morphology and evolution of urban space
Lect. 11. Urban morphology: Cozen School; Italian School; French School; Space Syntax Theory
Lect. 12. Physical environment of urban forms: urban forms and physical environment; environmental study; environmental evaluation and environmental operations
Lect. 13. Landscape urbanism: ideas, theories, operations and examples of landscape urbanism
Lect. 14. Researches on the phenomena of the urban self-organization: charms and problems of urban self-organization phenomena; research methodology on urban self-organization phenomena; case studies of urban self-organization phenomena
Lect. 15. Theory and method of architectural diagram: theoretical study on diagrams; concepts of architectural diagrams; application of diagram theory; diagrams as design tools; theoretical research of architectural diagrams in contemporary urban context
Lect. 16. Summary

Undergraduate Program 4th Year
LANDSCAPE PALNNING DESIGN AND THEORY • YIN Hang
Type: Elective Course
Study Period and Credits: 36 hours / 2 credits

Course Description
The object of landscape planning design includes all outdoor environments; the relationship between landscape and building is often close and interactive, which is especially obvious in a city. This course expects to carry out teaching from perspective of landscape design concept, site design technology, building's peripheral environment creation, etc., to establish a more comprehensive landscape knowledge system for the undergraduate students of architecture, and perfect their design ability in building site design, master plane planning and urban design and so on.
This course includes three aspects:
1. Concept and history
2. Site and context
3. Landscape and building

研究生一年级
景观都市主义理论与方法 · 华晓宁
课程类型：选修
学时/学分：18学时/1学分

Graduate Program 1st Year
THEORY AND METHOD OF LANDSCAPE URBANISM
· HUA Xiaoning
Type: Elective Course
Study Period and Credits: 18 hours / 1 credit

课程介绍
本课程作为国内首次以景观都市主义相关理论与策略为教学内容的尝试，介绍了景观都市主义思想产生的背景、缘起及其主要理论观点，并结合实例，重点分析了其在不同的场地和任务导向下发展起来的多样化的实践策略和操作性工具。
课程要求
　　1.要求学生了解景观都市主义思想产生的背景、缘起和主要理念。
　　2.要求学生能够初步运用景观都市主义的理念和方法分析和解决城市设计问题，从而在未来的城市设计实践中强化景观整合意识。
课程内容
　　第一讲：从图像到效能：景观都市实践的历史演进与当代视野
　　第二讲：生态效能导向的景观都市实践（一）
　　第三讲：生态效能导向的景观都市实践（二）
　　第四讲：社会效能导向的景观都市实践
　　第五讲：基础设施景观都市实践
　　第六讲：当代高密度城市中的地形学
　　第七讲：城市图绘与图解
　　第八讲：从原形到系统——AA景观都市主义

Course Description
Combining relevant theories and strategies of landscape urbanism firstly in China, the course introduces the background, origin and main theoretical viewpoint of landscape urbanism, and focuses on diversified practical strategies and operational tools developed under different orientations of site and task with examples.
Course Requirement
1. Students are required to understand the background, origin and main concept of landscape urbanism.
2. Students are required to preliminarily utilize the concept and method of landscape urbanism to analyze and solve the problem of urban design, so as to strengthen landscape integration consciousness in the future.
Course Content
Lecture 1: From pattern to performance: historical revolution and contemporary view of practice of landscape urbanism
Lecture 2: Eco-efficiency-oriented practice of landscape urbanism (1)
Lecture 3: Eco-efficiency-oriented practice of landscape urbanism (2)
Lecture 4: Social efficiency-oriented practice of landscape urbanism
Lecture 5: Infrastructure practice in landscape urbanism
Lecture 6: Geomorphology in contemporary high-density city
Lecture 7: Urban painting and diagrammatizing
Lecture 8: From prototype to system: aa landscape urbanism

研究生一年级
城市形态与设计方法论 · 丁沃沃
课程类型：必修
学时/学分：36学时/2学分

Graduate Program 1st Year
URBANISM & ARCHITECTURAL DESIGN METHODOLOGY·
DING Wowo
Type: Required Course
Study Period and Credits: 36 hours / 2 credits

课程介绍
　　建筑学核心理论包括建筑学的认识论和设计方法论两大部分。建筑设计方法论主要探讨设计的认知规律、形式的逻辑、形式语言类型，以及人的行为、环境特征和建筑材料等客观规律对形式语言的选择及形式逻辑的构成策略。为此，设立了提升建筑设计方法的关于设计方法论的理论课程，作为建筑设计及其理论硕士学位的核心课程。
课程要求
　　1.理解随着社会转型，城市建筑的基本概念在建筑学核心理论中的地位以及认知的视角。
　　2.通过理论的研读和案例分析理解建筑形式语言的成因和逻辑，并厘清中、西不同的发展脉络。
　　3.通过研究案例的解析理解建筑形式语言的操作并掌握设计研究的方法。
课程内容
　　第一讲：序言
　　第二讲：西方建筑学的基础
　　第三讲：中国：建筑的意义
　　第四讲：背景与文献研讨
　　第五讲：历史观与现代性
　　第六讲：现代城市形态演变与解析
　　第七讲：现代城市的"乌托邦"
　　第八讲：现代建筑的意义
　　第九讲：建筑形式的反思与探索
　　第十讲：建筑的量产与城市问题
　　第十一讲："乌托邦"的实践与反思
　　第十二讲：都市实践探索的理论价值
　　第十三讲：城市形态的研究
　　第十四讲：城市空间形态研究的方法
　　第十五讲：回归理性：建筑学方法论的新进展
　　第十六讲：建筑学与设计研究的意义
　　第十七讲：结语与研讨（一）
　　第十八讲：结语与研讨（二）

Course Description
Core theory of architecture includes epistemology and design methodology of Architecture. Architectural design methodology mainly discusses cognitive laws of design, logic of form and types of formal language, and the choice of formal language from objective laws such as human behavior, environmental feature and building material, and composition strategy of formal logic. Thus, the theory course about design methodology to promote architectural design methods is established as the core course of architectural design and theory master degree.
Course Requirement
1. To understand the status and cognitive perspective of basic concept of urban building in the core theory of architecture with the social transformation.
2. To understand the reason and logic of architectural formal language and different development process in China and West through reading theory and case analysis.
3. To understand the operation of architectural formal language and grasp methods of design study by analyzing study case.
Course Content
Lecture 1: Introduction
Lecture 2: Foundation of western architecture
Lecture 3: China: meaning of architecture
Lecture 4: Background and literature discussion
Lecture 5: Historicism and modernity
Lecture 6: Analysis and morphological evolution of modern city
Lecture 7: "Utopia" of modern city
Lecture 8: Meaning of modern Architecture
Lecture 9: Reflection and exploration of architectural form
Lecture 10: Mass production of buildings and urban problems
Lecture 11: Practice and reflection of "Utopia"
Lecture 12: Theoretical value of exploration on urban practice
Lecture 13: Study on urban morphology
Lecture 14: Method of urban spatial morphology study
Lecture 15: Return to rationality: new developments of methodology on architecture
Lecture 16: Meaning of architecture and design study
Lecture 17: Conclusion and discussion (1)
Lecture 18: Conclusion and discussion (2)

历史理论课程
HISTORY THEORY COURSES

本科二年级
中国建筑史（古代）· 赵辰
课程类型：必修
学时/学分：36学时/2学分

Undergraduate Program 2nd Year
HISTORY OF CHINESE ARCHITECTURE (ANCIENT)
• ZHAO Chen
Type: Required Course
Study Period and Credits: 36 hours / 2 credits

教学目标
本课程作为本科建筑学专业的历史与理论课程，目标在于培养学生的史学研究素养与对中国建筑及其历史的认识两个层面。在史学理论上，引导学生理解建筑史学这一交叉学科的多种棱面与视角，并从多种相关学科层面对学生进行基本史学研究方法的训练与指导。中国建筑史层面，培养学生对中国传统建筑的营造特征与文化背景建立构架性的认识体系。

课程内容
中国建筑史学七讲与方法论专题。七讲总体走向从微观向宏观，整体以建筑单体—建筑群体—聚落与城市—历史地理为序；从物质性到文化，建造技术—建造制度—建筑的日常性—纪念性—政治与宗教背景—美学追求。方法论专题包括建筑考古学、建筑技术史、人类学、美术史等层面。

Training Objective
As a mandatory historical & theoretical course for undergraduat students, this course aims at two aspects of training: the basi academic capability of historical research and the understandin of Chineses architectural history. It will help students to establish knowledge frame, that the discipline of History of Architecture as cross-discipline, is supported and enriched by multiple neighborin disciplines and that the features and development of Chines Architecture roots deeply in the natural and cultural background.

Course Content
The course composes seven 4-hour lectures on Chines Architecture and a series of lectures on methodology. The seven courses follow a route from individual to complex, from physica building to the intangible technique and to the cultural backgroun from technology to institution, to political and religious backgroun and finally to aesthetic pursuit. The special topics on methodolog include building archaeology, building science and technology anthropology, art history and so on.

本科二年级
外国建筑史（古代）· 王丹丹
课程类型：必修
学时/学分：36学时/2学分

Undergraduate Program 2nd Year
HISTORY OF WESTERN ARCHITECTURE (ANCIENT)
• WANG Dandan
Type: Required Course
Study Period and Credits: 36 hours / 2 credits

教学目标
本课程力图对西方建筑史的脉络做一个整体勾勒，使学生在掌握重要的建筑史知识点的同时，对西方建筑史在2000多年里的变迁的结构转折（不同风格的演变）有深入的理解。本课程希望学生对建筑史的发展与人类文明发展之间的密切关联有所认识。

课程内容
1. 概论 2. 希腊建筑 3. 罗马建筑 4. 中世纪建筑
5. 意大利的中世纪建筑 6. 文艺复兴 7. 巴洛克
8. 美国城市 9. 北欧浪漫主义 10. 加泰罗尼亚建筑
11. 先锋派 12. 德意志制造联盟与包豪斯
13. 苏维埃的建筑与城市 14. 1960年代的建筑
15. 1970年代的建筑 16. 答疑

Training Objective
This course seeks to give an overall outline of Wester architectural history, so that the students may have an in depth understanding of the structural transition (different styles of evolution) of Western architectural history in the past 2000 years. This course hopes that students can understand the clos association between the development of architectural history an the development of human civilization.

Course Content
1. Generality 2. Greek Architectures 3. Roman Architectures
4. The Middle Ages Architectures
5. The Middle Ages Architectures in Italy 6. Renaissance
7. Baroque 8. American Cities 9. Nordic Romanticism
10. Catalonian Architectures 11. Avant-Garde
12. German Manufacturing Alliance and Bauhaus
13. Soviet Architecture and Cities 14. 1960's Architectures
15. 1970's Architectures 16. Answer Questions

本科三年级
外国建筑史（当代）· 胡恒
课程类型：必修
学时/学分：36学时/2学分

Undergraduate Program 3rd Year
HISTORY OF WESTERN ARCHITECTURE (MODERN)
• HU Heng
Type: Required Course
Study Period and Credits: 36 hours / 2 credits

教学目标
本课程力图用专题的方式对文艺复兴时期的7位代表性的建筑师与5位现当代的重要建筑师作品做一细致的讲解。本课程将要重要建筑师的全部作品尽可能在课程中梳理一遍，使学生能够全面掌握重要建筑师的设计思想、理论主旨、与时代的特殊关联、在建筑史中的意义。

课程内容
1. 伯鲁乃列斯基 2. 阿尔伯蒂 3. 伯拉孟特
4. 米开朗琪罗（1） 5. 米开朗琪罗（2） 6. 罗马诺
7. 桑索维诺 8. 帕拉蒂奥（1） 9. 帕拉蒂奥（2）
10. 赖特 11. 密斯 12. 勒·柯布西耶（1）
13. 勒·柯布西耶（2） 14. 海杜克 15. 妹岛和世
16. 答疑

Training Objective
This course seeks to make a detailed explanation to the works of 7 representative architects in the Renaissance period and 5 important modern and contemporary architects in a special way. This course will try to reorganize all works of these importan architects, so that the students can fully grasp their design ideas, theoretical subject and their particular relevance with the era an significance in the architectural history.

Course Content
1. Brunelleschi 2. Alberti 3. Bramante
4. Michelangelo(1) 5. Michelangelo(2)
6. Romano 7. Sansovino 8. Palladio(1) 9. Palladio(2)
10. Wright 11. Mies 12. Le Corbusier(1) 13. Le Corbusier(2)
14. Hejduk 15. Kazuyo Sejima
16. Answer Questions

本科三年级
中国建筑史（近现代）· 赵辰 冷天
课程类型：必修
学时/学分：36学时/2学分

Undergraduate Program 3rd Year
HISTORY OF CHINESE ARCHITECTURE (MODERN)
• ZHAO Chen, LENG Tian
Type: Required Course
Study Period and Credits: 36 hours / 2 credits

课程介绍
本课程作为本科建筑学专业的历史与理论课程，是中国建筑史教学中的一部分。在中国与西方的古代建筑历史课程的基础上，了解中国社会进入近代，以至于现当代的发展进程。
在对比中西方建筑文化的基础之上，建立对中国近现代建筑的整体认识。深刻理解中国传统建筑文化在近代以来与西方建筑文化的冲突与相融之下，逐步演变发展至今天成为世界建筑文化的一部分之意义。

Course Description
As the history and theory course for undergraduate students of Architecture, this course is part of the teaching of History of Chinese Architecture. Based on the earlier studying of Chinese and Western history of ancient architecture, understand the evolution progress of Chinese society's entry into modern times and even contemporary age.
Based on the comparison of Chinese and Western building culture, establish the overall understanding of China's modern and contemporary buildings. Have further understanding of the significance of China's traditional building culture's gradual evolution into one part of today's world building culture under conflict and blending with Western building culture in modern times.

研究生一年级
建筑理论研究 · 王骏阳
课程类型：必修
学时/学分：18学时/1学分

Graduate Program 1st Year
STUDY OF ARCHITECTURAL THEORY · WANG Junyang
Type: Required Course
Study Period and Credits: 18 hours / 1 credit

课程介绍
　　本课程是西方建筑史研究生教学的一部分。主要涉及当代西方建筑界具有代表性的思想和理论，其主题包括历史主义、先锋建筑、批判理论、建构文化以及对当代城市的解读等。本课程大量运用图片资料，广泛涉及哲学、历史、艺术等领域，力求在西方文化发展的背景中呈现建筑思想和理论的相对独立性及关联性，理解建筑作为一种人类活动所具有的社会和文化意义，启发学生的理论思维和批判精神。

课程内容
　　第一讲 建筑理论概论
　　第二讲 数字化建筑与传统建筑学的分离与融合
　　第三讲 语言、图解、空间内容
　　第四讲 "拼贴城市"与城市的观念
　　第五讲 建构与营造
　　第六讲 手法主义与当代建筑
　　第七讲 从主线历史走向多元历史之后的思考
　　第八讲 讨论

Course Description
This course is a part of teaching Western architectural history for graduate students. It mainly deals with the representative thoughts and theories in Western architectural circles, including historicism, vanguard building, critical theory, tectonic culture and interpretation of contemporary cities etc.. Using a lot of pictures involving extensive fields including philosophy, history, art, etc., this course attempts to show the relative independence and relevance of architectural thoughts and theories under the development background of Western culture, understand the social and cultural significance owned by architectures as human activities, and inspire students' theoretical thinking and critical spirit.

Course Content
Lecture 1: Overview of architectural theories
Lecture 2: Separation and integration between digital architecture and traditional architecture
Lecture 3: Language, diagram and spatial content
Lecture 4: "Collage city" and concept of city
Lecture 5: Tectonics and Yingzao (Ying-Tsao)
Lecture 6: Mannerism and modern architecture
Lecture 7: Thinking after main-line history to diverse history
Lecture 8: Discussion

研究生一年级
建筑史方法 · 胡恒
课程类型：选修
学时/学分：18学时/1学分

Graduate Program 1st Year
THE STUDIES OF THE HISTORY OF ARCHITECTURE · HU Heng
Type: Elective Course
Study Period and Credits: 18 hours / 1 credit

教学目标
　　促进学生对历史研究的主题、方法、路径有初步的认识，通过具体的案例讲解使学生能够理解当代中国建筑史研究的诸多可能性。

课程内容
　　1.图像与建筑史研究（1-文学、装置、设计）
　　2.图像与建筑史研究（2-文学、装置、设计）
　　3.图像与建筑史研究（3-绘画与园林）
　　4.图像与建筑史研究（4-绘画、建筑、历史）
　　5.图像与建筑史研究（5-文学与空间转译）
　　6.方法讨论1
　　7.方法讨论2

Training Objective
To promote students' preliminary understanding of the topic, method and approach of historical research. To make students understand the possibilities of contemporary study on history of Chinese architecture through specific cases.

Course Content
1. Image and architectural history study (1-literature, device and design)
2. Image and architectural history study (2-literature, device and design)
3. Image and architectural history study (3- painting and garden)
4. Image and architectural history study (4- painting, architecture and history)
5. Image and architectural history study (5- literature and spatial transform)
6. Method discussion 1
7. Method discussion 2

本科二年级
CAAD理论与实践 · 童滋雨
课程类型：必修
学时/学分：36学时/2学分

Undergraduate Program 2nd Year
THEORY AND PRACTICE OF CAAD • TONG Ziyu
Type: Required Course
Study Period and Credits: 36 hours / 2 credits

课程介绍

在现阶段的CAD教学中，强调了建筑设计在建筑学教学中的主干地位，将计算机技术定位于绘图工具，本课程就是帮助学生可以尽快并且熟练地掌握如何利用计算机工具进行建筑设计的表达。课程中整合了CAD知识、建筑制图知识以及建筑表现知识，将传统CAD教学中教会学生用计算机绘图的模式向教会学生用计算机绘制有形式感的建筑图的模式转变，强调准确性和表现力作为评价CAD学习的两个最重要指标。

本课程的具体学习内容包括：
1. 初步掌握AutoCAD软件和SketchUP软件的使用，能够熟练完成二维制图和三维建模的操作；
2. 掌握建筑制图的相关知识，包括建筑投影的基本概念，平立剖面、轴测、透视和阴影的制图方法和技巧；
3. 图面效果表达的技巧，包括黑白线条图和彩色图纸的表达方法和排版方法。

Course Description

The core position of architectural design is emphasized in the CAD course. The computer technology is defined as drawing instrument. The course helps students learn how to make architectural presentation using computer fast and expertly. The knowledge of CAD, architectural drawing and architectural presentation are integrated into the course. The traditional mode of teaching students to draw in CAD course will be transformed into teaching students to draw architectural drawing with sense of form. The precision and expression will be emphasized as two most important factors to estimate the teaching effect of CAD course.
Contents of the course include:
1. Use AutoCAD and SketchUP to achieve the 2-D drawing and 3-D modeling expertly.
2. Learn relational knowledge of architectural drawing, including basic concepts of architectural projection, drawing methods and skills of plan, elevation, section, axonometry, perspective and shadow.
3. Skills of presentation, including the methods of expression and layout using mono and colorful drawings

---

本科三年级
建筑技术1——结构、构造与施工 · 傅筱
课程类型：必修
学时/学分：36学时/2学分

Undergraduate Program 3rd Year
ARCHITECTURAL TECHNOLOGY 1 — STRUCTURE, CONSTRUCTION AND EXECUTION • FU Xiao
Type: Required Course
Study Period and Credits:36 hours / 2 credits

课程介绍

本课程是建筑学专业本科生的专业主干课程。本课程的任务主要是以建筑师的工作性质为基础，讨论一个建筑生成过程中最基本的三大技术支撑（结构、构造、施工）的原理性知识要点，以及它们在建筑实践中的相互关系。

Course Description

The course is a major course for the undergraduate students of architecture. The main purpose of this course is based on the nature of the architect's work, to discuss the principle knowledge points of the basic three technical supports in the process of generating construction (structure, construction, execution), and their mutual relations in the architectural practice.

本科三年级
建筑技术2——建筑物理 · 吴蔚
课程类型：必修
学时/学分：36学时/2学分

Undergraduate Program 3rd Year
ARCHITECTURAL TECHNOLOGY 2 — BUILDING PHYSICS • WU Wei
Type: Required Course
Study Period and Credits:36 hours / 2 credits

课程介绍

本课程是针对三年级学生所设计，课程介绍了建筑热工学、建筑光学、建筑声学中的基本概念和基本原理，使学生掌握建筑的热环境、声环境、光环境的基本评估方法，以及相关的国家标准。完成学业后在此方向上能阅读相关书籍，具备在数字技术方法等相关资料的帮助下，完成一定的建筑节能设计的能力。

Course Description

Designed for the Grade 3 students, this course introduces the basic concepts and basic principles in architectural thermal engineering, architectural optics and architectural acoustics, so that the students can master the basic methods for the assessment of building's thermal environment, sound environment and light environment as well as the related national standards. After graduation, the students will be able to read the related books regarding these aspects, and have the ability to complete certain building energy efficiency designs with the help of the related digital techniques and methods.

本科三年级
建筑技术3——建筑设备 · 吴蔚
课程类型：必修
学时/学分：36学时/2学分

Undergraduate 3rd Year
ARCHITECTURAL TECHNOLOGY 3 — BUILDING EQUIPMENT • WU Wei
Type: Required Course
Study Period and Credits:36 hours / 2 credits

课程介绍

本课程是针对南京大学建筑与城市规划学院本科三年级学生所设计。课程介绍了建筑给水排水系统、采暖通风与空气调节系统、电气工程的基本理论、基本知识和基本技能，使学生能熟练地阅读水电、暖通工程图，熟悉水电及消防的设计、施工规范，了解燃气供应、安全用电及建筑防火、防雷的初步知识。

Course Description

This course is an undergraduate class offered in the School of Architecture and Urban Planning, Nanjing University. It introduces the basic principles of the building services systems, the technique of integration amongst the building services and the building. Throughout the course, the fundamental importance to energy, ventilation, air-conditioning and comfort in buildings are highlighted.

研究生一年级
传热学与计算流体力学基础 · 郜志
课程类型：选修
学时/学分：18学时/1学分

Graduate Program 1st Year
FUNDAMENTALS OF HEAT TRANSFER AND COMPUTATIONAL FLUID DYNAMICS • GAO Zhi
Type: Elective Course
Study Period and Credits: 18 hours / 1 credit

课程介绍

本课程的主要任务是使建筑学/建筑技术学专业的学生掌握传热学和计算流体力学的基本概念和基础知识，通过课程教学，使学生熟悉传热学中导热、对流和辐射的经典理论，并了解传热学和计算流体力学的实际应用和最新研究进展，为建筑能源和环境系统的计算和模拟打下坚实的理论基础。教学中尽量简化传热学和计算流体力学经典课程中复杂公式的推导过程，而着重于如何解决建筑能源与建筑环境中涉及流体流动和传热的实际应用问题。

Course Description

This course introduces students majoring in building science and engineering / building technology to the fundamentals of heat transfer and computational fluid dynamics (CFD). Students will study classical theories of conduction, convection and radiation heat transfers, and learn advanced research developments of heat transfer and CFD. The complex mathematics and physics equations are not emphasized. It is desirable that for real-case scenarios students will have the ability to analyze flow and heat transfer phenomena in building energy and environment systems.

研究生一年级
GIS基础与应用 · 童滋雨
课程类型：选修
学时/学分：18学时/1学分

Graduate Program 1st Year
CONCEPT AND APPLICATION OF GIS • TONG Ziyu
Type: Elective Course
Study Period and Credits:18 hours / 1 credit

课程介绍

本课程的主要目的是让学生理解GIS的相关概念以及GIS对城市研究的意义，并能够利用GIS软件对城市进行分析和研究。

Course Description

This course aims to enable students to understand the related concepts of GIS and the significance of GIS to urban research, and to be able to use GIS software to carry out urban analysis and research.

研究生一年级
建筑环境学・邵志
课程类型：选修
学时/学分：18学时 / 1学分

Graduate Program 1st Year
FUNDAMENTALS OF BUILT ENVIRONMENT • GAO Zhi
Type: Elective Course
Study Period and Credits:18 hours / 1 credit

课程介绍

本课程的主要任务是使建筑学/建筑技术学专业的学生掌握建筑环境的基本概念，学习建筑与城市热湿环境、风环境和空气质量的基础知识。通过课程教学，使学生熟悉城市微气候等理论，并了解人体对热湿环境的反应，掌握建筑环境学的实际应用和最新研究进展，为建筑能源和环境系统的测量与模拟打下坚实的基础。

Course Description

This course introduces students majoring in building science and engineering / building technology to the fundamentals of built environment. Students will study classical theories of built / urban thermal and humid environment, wind environment and air quality. Students will also familiarize urban micro environment and human reactions to thermal and humid environment. It is desirable that students will have the ability to measure and simulate building energy and environment systems based upon the knowledge of the latest development of the study of built environment.

研究生一年级
材料与建造・冯金龙
课程类型：必修
学时/学分：18学时 / 1学分

Graduate Program 1st Year
MATERIAL AND CONSTRUCTION • FENG Jinlong
Type: Required Course
Study Period and Credits:18 hours / 1 credit

课程介绍

本课程将介绍现代建筑技术的发展过程，论述现代建筑技术及其美学观念对建筑设计的重要作用；探讨由材料、结构和构造方式所形成的建筑建造的逻辑方式；研究建筑形式产生的物质技术基础，诠释现代建筑的建构理论与研究方法。

Course Description

It introduces the development process of modern architecture technology and discusses the important role played by the modern architecture technology and its aesthetic concepts in the architectural design. It explores the logical methods of construction of the architecture formed by materials, structure and construction. It studies the material and technical basis for the creation of architectural form, and interprets construction theory and research methods for modern architectures.

研究生一年级
计算机辅助建筑设计技术・吉国华
课程类型：必修
学时/学分：36学时 / 2学分

Graduate Program 1st Year
COMPUTER AIDED ARCHITECTURAL DESIGN • JI Guohua
Type: Elective Course
Study Period and Credits:36 hours / 2 credits

课程介绍

随着计算机辅助建筑设计技术的快速发展，当前数字技术在建筑设计中的角色逐渐从辅助绘图转向了真正的辅助设计，并引发了设计的革命和建筑的形式创新。本课程讲授Grasshopper参数化编程建模方法以及相关的几何知识，让学生在掌握参数化编程建模技术的同时，增强以理性的过程思维方式分析和解决设计问题的能力，为数字建筑设计和数字建造打下必要的基础。

基于Rhinoceros的算法编程平台Grasshopper的参数化建模方法，讲授内容包括各类运算器的功能与使用、图形的生成与分析、数据的结构与组织、各类建模的思路与方法，以及相应的数学与计算机知识。

Course Description

The course introduces methods of Grasshopper parametric programming and modeling and relevant geometric knowledge. The course allows students to master these methods, and enhance ability to analyze and solve designing problems with rational thinking at the same time, building necessary foundation for digital architecture design and digital construction.

In this course, the teacher will teach parametric modeling methods based on Grasshopper, a algorithmic programming platform for Rhinoceros, including functions and application of all kinds of arithmetic units, pattern formation and analysis, structure and organization of data, various thoughts and methods of modeling, and related knowledge of mathematics and computer programming.

研究生一年级
建筑体系整合・吴蔚
课程类型：选修
学时/学分：18~36学时 / 1~2学分

Graduate Program 1st Year
BUILDING SYSYTEM INTEGRATION • WU Wei
Type: Elective Course
Study Period and Credits: 18~36 hours / 1~2 credits

课程介绍

本课程是从建筑各个体系整合的角度来解析建筑设计。首先，课程介绍了建筑体系整合的基本概念、原理及其美学观念；然后具体解读以上各个设计元素在整个建筑体系中所扮演的角色及其影响力，了解建筑各个系统之间的相互联系和作用；最后，以全球的环境问题和人类生存与发展为着眼点，引导同学们重新审视和评判我们奉为信条的设计理念和价值体系。本课程着重强调建筑设计需要了解不同学科和领域的知识，熟悉各工种之间的配合和协调。

Course Description

A building is an assemblage of materials and components to obtain a shelter from external environment with a certain amount of safety so as to provide a suitable internal environment for physiological and psychological comfort in an economical manner. This course examines the role of building technology in architectural design, shows how environmental concerns have shaped the nature of buildings, and takes a holistic view to understand the integration of different building systems. It employs total building performance which is a systematic approach, to evaluate the performance of various sub-systems and to appraise the degree of integration of the sub-systems.

回声 ECHO

乡土建造与建构访谈会
RURAL BUILDING AND CONSTRUCTION INTERVIEW

目标
以南大建筑实践为根基，从乡土建造与建构开始，不断探讨中国未来建筑实践的方向和方法，让南京大学建筑同仁有集体发声之处，让南大建筑形成合力。

研讨主题
材料结构探索；建造教学、建造实验；乡村复兴案例

参会人员
王磊（召集人）、王铠、孟凡浩、张东光、罗辉、杨保新、毕胜、唐涛、吴子夜、周超、孙久强、李亚伟、杜春宇

负责人
周凌、王铠、赵辰、王丹丹

发言摘录
王铠：
演讲主题为当代乡土的原生秩序。

当下中国建筑业界的"乡建"，作为政府主导的"乡村振兴"浩大社会工程的积极有生力量，正是面对传统聚落价值自发和自觉的探索。乡村的问题是我国重新构建传统农业文明与现代工业文明、城市文明关系的核心，体现着多方面社会力量的共同作用：政府对传统农业文明现代转型的预期与制度建设；资本投入对乡村经济发展的产业推动；知识分子对乡土文化的认知与思考；最重要的是农村广大人民对乡土生活改善的内在需求。

毕胜：
演讲主题为乡村建设中如何留住特有的地域情怀——对比日本合掌村的改造历程浅谈莱州初家村。

目前，中国的乡村建设正如火如荼地在各地展开，这里我们以所参加的莱州初家村的改造为切入点，通过与日本白川乡合掌村的改造和发展历程的对比研究，探讨初家村改造与发展的可行性方式，借此为中国乡村建设的研究、保护、开发提供有益的借鉴。

罗辉：
演讲主题为闽东乡村复兴。

在当下以自然/人文生态保护下的乡村发展为目标的乡村复兴中，乡村聚落与民居空间的生态发展成为当下建筑学理论研究和实践活动之核心。拨开"形式风格"之迷雾回到传统建造体系层面，认知其空间上和时间上对自然地理和社会文化条件的应变性，从中辨析现代化路径并实践之，是乡村空间得以生态发展的关键。

李亚伟：
演讲主题为乡村休憩空间营造两则。

其实建筑设计本身不是重点，而是证明一个观点。设计之初，有一个针对当时新农村建设和城镇化过程中的疑问：是迎合当地审美保持地域特色：白墙灰瓦，还是简单质朴、构造真实的现代设计？以最终的结果和使用者的评价来看，得到一个基本结论：真正的乡建其实是当地材料、建造、形式等建筑本体的研究与思考就好，不用刻意迎合风格与样式。地域特色和表达其实就隐藏在基地、气候、材料、习惯之中。简洁美好的空间可以融入当地环境，融入当地生活。

孟凡浩：
演讲主题为现实与理想：近期乡村实践的思考。

短短数十年间，随着生活方式的多样化和营造物质类型手段的日益丰富，城市建筑的象征意义被不断夸大，乡村建筑的内在建构逻辑也被城市化侵蚀所湮没。而建筑师的主要职责却被认定为物质生产者——创造并追求物质的精良。

随着城市营建和乡村激活领域的不断深入，城市和乡村已成为不可分割的社会生态体系，需要建筑师以更宏观的视角介入与尝试。也正是因为如此，在城乡一体化的大背景下，我们认为，任何脱离城市谈乡村或不管乡村只重城市的社会实践，都难以取得最终平衡。

驿道廊桥改造 Post corridor bridge reconstruction

唐涛：
演讲主题为农庄生态大棚的改造实践。
建筑师在项目中始终需要反问自己的三个问题。（1）项目的真正需求以及所面临的真正问题是什么。（2）建筑师在项目中究竟需要扮演什么样的角色。（3）建筑师多年的专业训练与能力通过怎样的方式能够真正体现它的价值。这些反思是帮助我们拨开云雾，找到本质的必经过程。

王磊：
演讲主题为乡村系统营造之法。
"以农民为主体的陪伴式系统乡建"是我的乡村实践的核心。系统性地切入乡村工作，选择了一个地方就永远陪伴下去，只是当好协作者。乡村实践最重要的工作是把农民组织起来，让村集体资源整合，建立和壮大村集体经济，并且可持续增长；协助村集体提高服务村民的能力，让更多的村民脱贫，实现共同富裕。实践目的是实现"生产、生活、生态"三生共赢的新农村，用社区营造的方法进行地域性建造，自然而然地设计，和农民一起建设和改造乡村。

吴子夜：
演讲主题为蒋山渔村更新实验。
建筑师在乡村实践中的参与程度或有不同，但是无论是规划、制度，还是功能、空间和建造，都应该从乡村本源的"人"的角度出发，以村民最质朴的生活和文化需求为思考起点，以在地文化的现代表达为手段，影响并复兴乡村。

张东光：
演讲主题为基于建构的整合性设计。
建筑师进入乡村工作，如果是在行政意愿或资本逐利的背景下，需要清楚地辨析项目的各种外部条件以及自身的工作边界。大规模及快速的规划工作可能会给原本自然发展的乡村带来伤害。因此，在不成熟的条件下，笔者更倾向于建筑师轻微地介入其中。在专业领域内，通过村庄整体环境、风貌的提升以及用个别创新作品来引导村庄发展之外，建筑师还可以选择去解决一些居住生活品质的基本问题。致力于原型的开拓和发展，也许会带来更大的社会意义和价值。

杜春宇：
演讲主题为浙江模式乡村振兴——田园综合体及特色小镇的传承与创新。
随着新村的建成使用，当地人的意识也在逐渐转变，原来不是只有模仿西式洋房才是富裕的标志，桃花源一样的中国田园式农居同样是村民真正向往的生活方式。周边的村民纷纷来要图纸，在老村自建房摸索建设，这是我们建筑师最愿意看到的，设计成果被村民认可，引导了新的审美及生活方式，并被老百姓自发地借鉴、推广。

周超：
演讲主题为轻型建筑的乡村实践。
建筑师在乡村实践，首先要立足于乡村的生态资源和环境保护，尊重当地村落的原生秩序，熟悉当地的人文传统，了解产业经济的发展状况。应根据当地的地貌肌理、气候特征和建造传统，选择合适的建造手段，并遵循地域和建构原则。我们在乡村实践中，尝试以轻型建筑介入部分建造活动，利用工业化的建造系统，包括胶合木、胶合竹和轻钢等材料，在工厂里预制构件，在现场装配化施工。这种建造方式，既充分发挥了轻型系统对土地资源的较少破坏、节约材料、建造速度较快等优势，又与传统建造方法互为补充，实现乡土生活的当代演绎。

孙久强：
演讲主题为乡土空间营造的适用策略。
从现代主义建筑思想出发构建基本设计方法，真实地挖掘乡村社会的空间需求，以地域材料和本土技术营造生动的空间体验，以建筑和空间为载体探寻乡村建设的"适用之道"。

杭州富阳东梓关回迁农居 Dongziguan Affordable Housing for Relocalized Farmers

杨保新：
演讲主题为当代竹空间于乡建语境中的可能性。

竹在大尺度的"竹空间"层面一直以来发展水平不高，缺少持续的探索和创新，多呈现两种现象，一是民间仿木的竹楼和竹亭，二是以竹作为装饰的"竹装修"。前者，竹压抑于木的形式下没能表现出自身特性；后者，竹仅作为外装饰，并不构成真正意义上的竹空间。当代"竹空间"应是以竹作为主要语言，将"材料—结构—空间—肌理"整合于一体，并呈现出"竹"这一元素独特气质的空间形态。

Objective
Base on the architectural practice of School of Architecture, Nanjing University, start with rural building and construction, constantly discuss future direction and method of architectural practice in China, make the architectural colleagues of Nanjing University give opinion, and form synergy.

Discussion Topic
Material structure exploration; building teaching, building experiment; rural revitalization case

Participants
WANG Lei (convener), WANG Kai, MENG Fanhao, ZHANG Dongguang, LUO Hui, YANG Baoxin, BI Sheng, TANG Tao, WU Ziye, ZHOU Chao, SUN Jiuqiang, LI Yawei, DU Chunyu

Principals
ZHOU Ling, WANG Kai, ZHAO Chen, WANG Dandan

WANG Kai:
Speech theme is "Original order of contemporary countryside".
Current "rural construction" in Chinese architecture industry, as the positive and effective strength for "rural revitalization" social program led by the government, is the spontaneous and conscious exploration of traditional settlement value. Rural problem is the core for China to reconstruct relationship between traditional agricultural civilization and modern industrial civilization, urban civilization, and reflects the joint function of social forces: expectation of the government on modern transformation of traditional agricultural civilization and institutional construction; industrial promotion from capital input on rural economic development; cognition and thinking from intellectuals on rural culture; the most important is the internal need of rural people on improving rural life.

BI Sheng:
Speech theme is "How to retain unique regional emotion in rural construction: brief talk of Chujia Village in Laizhou through comparison with the reform of Shirakawa Village in Japan".
At present, the rural construction in China is growing vigorously, we start with the reform of Chujia Village in Laizhou that we participate in, through comparison with the reform and development of Shirakawa Village in Japan, discuss the feasible method for reform and development of Chujia Village, so as to provide beneficial reference for research, protection, development of Chinese rural construction.

LUO Hui:
Speech theme is "Rural revitalization in east Fujian".
In current rural revitalization that aims for rural development under natural/cultural ecological protection, ecological development of rural settlement and dwelling space becomes the core of theoretical research and practical activity of current architecture. Expelling the mist of "formal style" and return to the level of traditional construction system, cognizing the adaptability to natural geography and social cultural condition on space and time, and analyzing and practicing modernization path is the key point for ecological development of rural space.

武汉市江夏区五里界街道童周岭村小朱湾 Xiaozhu Bay in Tongzhouling, Wuhan

LI Yawei:
Speech theme is "Two examples of rural rest space construction".
In fact, architectural design is not a focus but to prove a view. At the beginning of design, there is a question for current new rural construction and urbanization: to cater local aesthetics and retain regional characteristics: white wall and grey tile, or to be simple modern design with real construct? From the final result and evaluation of user, a basic conclusion is obtained: the real rural construction is actually based on the research and thinking on local material, construction, form, etc. instead of deliberately catering style and type. Regional characteristics and expression are hidden in foundation, climate, material, habit. Brief and beautiful space can be merged in local environment and local life.

MENG Fanhao:
Speech theme is "Reality and ideal: thinking of recent rural practice".
In a few decades, with the diversification of lifestyle and increasing enriched means of building material, the symbolic meaning of urban architecture is constantly exaggerated, and the internal construction logic of rural architecture is also eroded by urbanization. The main duty of architect is considered to be material producer—creating and pursuing improvement of material.
With constant progress of urban building and rural activated domain, urban area and rural area have become inseparable social ecological system, which requires the architect to get involved and try from broader view. Therefore, under the background of urban-rural integration, we believe that it will be difficult to obtain the final balance if we only talk rural area without urban area, or only focus on urban social practice despite of rural area.

TANG Tao:
Speech theme is "Reform practice of farm ecological greenhouse".

Architect should always rethink three questions in the project. Firstly, what is the real project requirement and real problem? Secondly, what is the role the architect should play in the project? Thirdly, how to really reflect the value of years of professional training and ability of the architect? This can help us find the essentials.

WANG Lei:
The theme of the speech is "The method of systematic rural construction".
"The systematic farmer-oriented companion rural construction" is the core of my rural practice. Carry out the rural work in a systematic manner, always stay with the rural construction once choosing a place, and be a good collaborator. The most important work of rural practice is to organize farmers to integrate the village collective resources and establish and strengthen the village collective economy and sustainable development; assist the village collective to improve the ability to serve the villagers and enable more villagers to shake off poverty and achieve common prosperity. The purpose of the practice is to establish a new village with the mutual benefits of "production, life and ecology". Implement regional construction and natural design in the community-based approach, and build and transform the village with the farmers.

WU Ziye:
Speech theme is "Jiangshan Fishing Village renewal experiment".
The degree of participation in rural practice is different, however, no matter it is planning, system, or function, space and construction, architect should start with the perspective of "human", the origin of countryside, think about the simplest living and cultural needs of villagers, by means of modern expression of local culture, affect and realize rural revitalization.

ZHANG Dongguang:
Speech theme is "Integration design based on tectonic theory".

浙江余姚中村竹桥 Yuyao Nakamura bamboo bridge landscape

When entering the countryside to work, if for administrative will or pursuing profit, architect should clearly identify various external conditions of the project and own working boundary. Large-scale and rapid planning work may damage the natural development of countryside. Therefore, in immature condition, the writer prefers to minor intervention of architect. In the professional field, through improvement of overall rural environment, appearance, and using individual innovative works to guide rural development, architect can also choose to solve some basic problem of living quality. Being committed to development of prototype may bring with greater social meaning and value.

DU Chunyu:
Speech theme is "Rural revitalization in Zhejiang mode: inheritance and innovation of countryside complex and characteristic town".
With the completion and use of new village, the awareness of locals is gradually transforming, not only simulating western style house is the symbol of richness, the paradise-like Chinese rural village is the lifestyle that villagers really long for. Surrounding villagers ask for drawing and explore to construct in the old village, this is the scene that architects love to see: the design result is recognized by the villagers, guides their new aesthetics and lifestyle, and is learned, promoted by common people voluntarily.

Zhou Chao:
Speech theme is "Rural practice of light building".
To practice in countryside, architects should base on rural ecological resource and environmental protection, respect original order of local village, get familiar with local culture and tradition, and understand development condition of industrial economy. Architects should choose proper construction method according to local landform texture, climatic characteristics and construction tradition, and abide by regional and construction principle. In the rural practice, we try to use light building to intervene in some construction activity, utilize industrialized construction system, including laminated wood, laminated bamboo and light steel, etc.. to prefabricate component in the factory, and assemble on the site. Such construction method does not only give full play to the advantages of light system such as less damage on the land resource, saving material, fast construction, etc., but also complements traditional construction method, and realizes contemporary rural life.

Sun Jiuqiang:
Speech theme is "Policy applicable to rural space building".
Construct basic design methods from modernism architectural thought, really excavate space requirement of rural society, use regional material and local technique to build vivid space experience, and search for "applicable way" for rural construction on the carrier of architecture and space.

Yang Baoxin:
Speech theme is "Possibility of contemporary bamboo space in rural construction context".
The development level of bamboo on the level of large-scale "bamboo space" lack of continuous exploration and innovation, and is mainly reflected in two appearances, one is folk wood-like bamboo tower and bamboo kiosk, another is "bamboo decoration" decorated by bamboo. In the former one, bamboo is suppressed by wooden form without showing its own characteristics; in the latter one, bamboo is only used as exterior decoration without composing real bamboo space. Contemporary "bamboo space" should take bamboo as the main language, integrate "material, structure, space, texture", and present the unique space form of element "bamboo".

蒋山渔村更新实践 Jiangshan Fishing Village Renewal

吴子夜：

改造部分以蒋山书舍为例，设计从村落中功能的缺失开始思考，强调逻辑的完整性。通过打破建筑功能、空间和体验的界限，给村民们带来新的活动场所，并影响他们对传统老宅的固有看法。同时结合建筑原貌与后续的使用对比进行反思，就改造类项目的尺度和新老平衡的关系展开讨论。

Wu Ziye:
The reform part takes Jiangshan book shed for an example, the design starts with the loss of function in village, and emphasizes on the integrity of logic. Through breaking the boundary of building function, space and experience, we bring the villagers new activity place, and affect their inherent opinion of traditional old house. At the same time, we rethink the original building appearance and comparison with subsequent use, and discuss on the scale of reform project and balance between new and old.

南院 The south yard in Guangxi

周超:"竹钢"是一种新型竹纤维高强复合材料,它除了具有普通竹材的速生、环保、节能等特点外,还因其特殊的生产工艺,使得该材料的力学性能远远超过其他竹材,更重要的是,它可以作为结构材料来使用。胶合竹预制建筑广泛应用于临时建筑、景区建筑、更新与改造等。

ZHOU Chao: "Bamboo steel" is a new type bamboo fiber high-strength composite material, besides the rapid growth, environmental, energy saving characteristics of ordinary bamboo material, it also has special production process, which makes its mechanical property far better then other bamboo material, moreover, it can be used as structure material. Laminated bamboo prefabricated building is widely applied in temporary building, scenic area building, renewal and reform, etc..

遵义桐梓县内置金融村社及联合社
The finance association and community built-in Tongzi county, Zunyi

南院 The south yard in Guangxi

河北阜平龙泉关森林驿站 Longquanguan Forest Station

讲座 Lectures

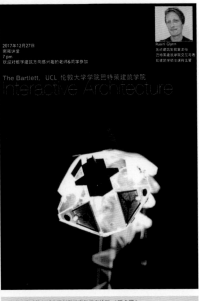

硕士学位论文列表
List of Thesis for Master Degree

研究生姓名	研究生论文标题	导师姓名
王晓茜	古城风貌控制下的精品酒店设计研究——以丽江古城文治巷42号院落改造项目为例	张 雷
余星凯	南京印染厂旧址再利用改造设计研究	张 雷
张 学	景德镇丙丁柴窑窑房研究及设计	张 雷
冯 琪	基于少年儿童行为的小学校园设计研究	张 雷
席 弘	中国新乡土建筑营造的适宜技术研究——以江宁石塘村互联网会议中心为例	张 雷
张 欣	基于当代城市青年行为特征与居住需求的青年公寓设计研究	张 雷
陈立华	南京师范大学中北学院体育馆建筑方案设计	冯金龙
蒋建昕	南师大中北学院多功能剧院设计研究	冯金龙
谢忠雄	江苏省农科院科技园建筑设计	冯金龙
赵 伟	江苏省地址资料库建筑设计	冯金龙
王子珊	度假酒店聚落型空间设计研究	冯金龙
吴松霖	原竹建筑结构性节点研究及其设计表达	冯金龙
黄凯峰	互承结构的形式生成研究	吉国华
江振彦	极小曲面的参数化生成与设计	吉国华
施 成	南京大学鼓楼校区自行车棚设计——参数化在设计中的简单运用	吉国华
张本纪	参数化砖墙的生形设计与失稳计算研究	吉国华
贾福龙	基于Miura-Ori的曲面褶皱化造型研究	吉国华
王峥涛	基于可展开面的造型方法研究	吉国华
杨肇伦	曲面六边形平面化重构研究	吉国华
宋富敏	徐家院文旅中心综合楼及民宿酒店设计	周 凌
张豪杰	南京徐家院村民宿改造设计及研究	周 凌
周贤春	老年公寓模块产业化设计研究	周 凌
邵思宇	元阳哈尼梯田遗产区传统村落人居环境修复研究	周 凌
拓 展	低造价本土适宜性技术的应用研究——以南京苏家村改造为例	周 凌
陈嘉铮	常州武进区湖塘老街保护规划建筑设计方案——北街A街块	傅 筱
刘泽超	常州武进区湖塘老街保护规划建筑设计方案——北街C街块	傅 筱
吕秉田	常州武进区湖塘老街保护规划建筑设计方案——北街B街块	傅 筱
吴结松	金坛茅山游客服务中心设计	傅 筱
李文凯	幼儿园建筑安全性构造设计研究	傅 筱
徐一品	沿海经济发达地区大量性建筑围护结构回应气候状况调查研究	傅 筱

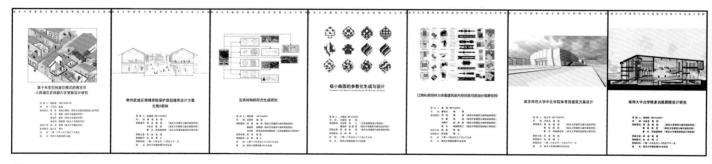

研究生姓名	研究生论文标题	导师姓名
缪姣姣	基于共享空间居住模式的南京小西湖历史风貌片区更新设计研究	丁沃沃
邹晓蕾	城市历史片区居住空间更新设计研究——以南京市小西湖片区为例	丁沃沃
顾聿笙	城市街廓空间风环境模拟试验区范围设定研究	丁沃沃
李若尧	基于风环境质量的城市住区更新潜力分析	丁沃沃
杨益晖	居住区建筑室外空间与风环境关联性研究	丁沃沃
刘晓君	霍华德·阿克利的艺术之家	华晓宁、保罗·沃克
宋春亚	传统村落复兴中的公共功能植入研究——以高修村村落中心复兴为例	华晓宁
周明辉	水位大落差条件下滨水空间设计研究——高淳区武家嘴村蛇山片区滨水区域改造设计	华晓宁
胡珊	建造体系视角下的山地民居聚落空间研究——以广西柳州市为例	华晓宁
于慧颖	基于当代城市公共休闲需求的近代建筑文化遗产保护与再利用设计——以原中央体育场旧址国术场为例	赵辰
刘垄鑫	交通因素主导下的慈城古县城边缘带形态研究	赵辰
曹阳	中国建筑学语境下的现代建筑话语研究（1950s）	王骏阳
赵婧靓	水晶宫的环境调控与现代建筑的技术维度	王骏阳
姜澜	江南私家园林大体量建筑室内空间非均质划分现象初探	鲁安东
张黎萌	中国城市形态学理论中对于城市扩张认识的演变研究——以"生长机制"术语概念为线索	鲁安东
周洋	结合互动技术的纪念性空间设计研究——以侵华日军南京大屠杀遇难同胞纪念馆冥思厅改造为例	鲁安东
崔傲寒	建筑设计视角下记忆场所营造框架初探	鲁安东
方飞	民国镇江地区蚕种场的建设活动研究	鲁安东
王却奁	勒·柯布西耶的电影术——以空间的电影性特征为视角对柯布西耶建筑设计的初步研究	鲁安东
艾心	小学建筑空间组织研究——以宝华山小学建筑设计为例	童滋雨
种桂梅	基于微气候效应的城市多层居住区内开放空间优化配置研究	童滋雨
蒋靖才	南京地区医院建筑视觉与非视觉效应的天然采光研究	吴蔚
程思远	夏热冬冷地区住宅建筑新风热回收系统节能效果研究	秦孟昊
黄丽	影响家具木器漆挥发性有机化合物（VOCs）散发关键因素研究	秦孟昊
王敏姣	家用纺织品的湿缓冲现象对室内热湿环境的影响	秦孟昊
沈佳磊	室内臭氧的来源、去除与分布特性研究	郗志
张洪光	机器学习方法在室内污染源特性识别的研究	郗志
彭丹丹	夏热冬冷地区现代住区肌理的微气候性能研究——以南京夏季时段为例	胡友培、赵辰
张靖	普通村镇多层砖混建筑改造与再利用设计研究——以江阴市周庄镇滨河路某建筑为例	胡友培、吉国华
柳纬宇	历史风貌地段城市肌理形态控制导则城市设计研究——以忠义街地段为例	胡友培、丁沃沃

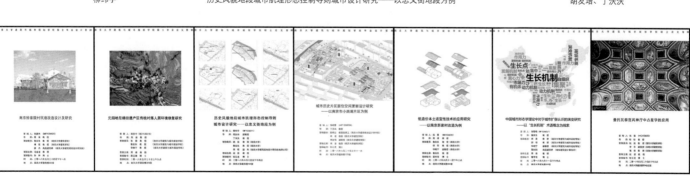

在校学生名单
List of Students

本科生 Undergraduate

2014级学生 / Students 2014

蔡英杰 CAI Yingjie	林 宇 LIN Yu	施孝萱 SHI Xiaoxuan	谢 峰 XIE Feng
曹 焱 CAO Yan	刘 畅 LIU Chang	宋 怡 SONG Yi	严紫微 YAN Ziwei
陈妍霓 CHEN Yanni	刘宛莹 LIU Wanying	宋宇宁 SONG Yuning	杨云睿 YANG Yunrui
杜孟泽杉 DU Mengzeshan	刘为尚 LIU Weishang	宋云龙 SONG Yunlong	杨 钊 YANG Zhao
胡皓捷 HU Haojie	卢 鼎 LU Ding	唐 萌 TANG Meng	尹子晗 YIN Zihan
兰 阳 LAN Yang	马西伯 MA Xibo	完颜尚文 WANYAN Shangwen	张 俊 ZHANG Jun
梁晓蕊 LIANG Xiaorui	施少鋆 SHI Shaojun	夏心雨 XIA Xinyu	张珊珊 ZHANG Shanshan

2015级学生 / Students 2015

卞秋怡 BIAN Qiuyi	顾梦婕 GU Mengjie	刘 越 LIU Yue	汪 榕 WANG Rong	杨 洋 YANG Yang
陈景杨 CHEN Jingyang	何 璇 HE Xuan	罗逍遥 LUO Xiaoyao	王 晨 WANG Chen	叶庆锋 YE Qingfeng
戴添趣 DAI Tianqu	兰贤元 LAN Xianyuan	罗紫璇 LUO Zixuan	王雪梅 WANG Xuemei	张昊阳 ZHANG Haoyang
邸晓宇 DI Xiaoyu	李博文 Li Bowen	吕文倩 LV Wenqian	卫 斌 WEI Bin	赵 彤 ZHAO Tong
丁展图 DING Zhantu	李心仪 LI Xinyi	毛志敏 MAO Zhimin	仙海斌 XIAN Haibin	周 杰 ZHOU Jie
龚 正 GONG Zheng	刘 博 LIU Bo	秦伟航 QIN Weihang	徐玲丽 XU Lingli	
顾卓琳 GU Zhuolin	刘秀秀 LIU Xiuxiu	沈静雯 SHEN Jingwen	杨鑫毓 YANG Xinyu	

2016级学生 / Students 2016

陈 帆 CHEN Fan	龚之璇 GONG Zhixuan	潘 博 PAN Bo	谢琳娜 XIE Linna
陈鸿帆 CHEN Hongfan	黄靖绮 HUANG Jingqi	丘雨辰 QIU Yuchen	于文爽 YU Wenshuang
陈婧秋 CHEN Jingqiu	黄文凯 HUANG Wenkai	石雪松 SHI Xuesong	余沁蔓 YU Qinman
陈铭行 CHEN Mingxing	雷 畅 LEI Chang	司昌尧 SI Changyao	张涵筱 ZHANG Hanxiao
陈应楠 CHEN Yingnan	李宏健 LI Hongjian	王 路 WANG Lu	周子琳 ZHOU Zilin
陈予婧 CHEN Yujing	李舟涵 LI Zhouhan	吴林天池 WU Lintianchi	朱凌云 ZHU Lingyun
封 翘 FENG Qiao	马彩霞 MA Caixia	吴敏婷 WU Minting	

2017级学生 / Students 2017

卞直瑞 BIAN Zhirui	樊力立 FAN Lili	刘 畅 LIU Chang	沈葛梦欣 SHEN Gemengxin	张凯莉 ZHANG Kaili
卜子睿 BU Zirui	甘静雯 GAN Jingwen	龙 沄 LONG Yun	沈晓燕 SHEN Xiaoyan	周金雨 ZHOU Jinyu
陈佳晨 CHEN Jiachen	顾天奕 GU Tianyi	陆柚余 LU Youyu	孙 萌 SUN Meng	周慕尧 ZHOU Muyao
陈露茜 CHEN Luxi	韩小如 HAN Xiaoru	马路遥 MA Luyao	孙 瀚 SUN Han	朱菁菁 ZHU Jingjing
陈雨涵 CHEN Yuhan	焦梦雅 JIAO Mengya	马子昂 MA Ziang	杨 帆 YANG Fan	朱雅芝 ZHU Yazhi
程科懿 CHENG Keyi	李心彤 LI Xintong	彭 洋 PENG Yang	杨佳锟 YANG Jiakun	达热亚·阿吾斯哈力 Dareya AWUSIHALI
董一凡 DONG Yifan	林易谕 LIN Yiyu	尚紫鹏 SHANG Zipeng	杨乙彬 YANG Yibin	

研究生 Postgraduate

车俊颖 CHE Junying	顾一蝶 GU Yidie	梁万富 LIANG Wanfu	刘文沛 LIU Wenpei	孙雅贤 SUN Yaxian	王政 WANG Zheng	徐麟 XU Lin	杨玉茵 YANG Yuhan	张丛 ZHANG Cong
陈博宇 CHEN Boyu	韩书园 HAN Shuyuan	梁耀波 LIANG Yaobo	刘宇 LIU Yu	谭健 TAN Jian	武春洋 WU Chunyang	徐思恒 XU Siheng	姚晨阳 YAO Chenyang	张海宁 ZHANG Haining
陈凌杰 CHEN Lingjie	胡任元 HU Renyuan	林琛 LIN Chen	陆扬帆 LU Yangfan	田金华 TIAN Jinhua	吴昇奕 WU Shengyi	徐天驹 XU Tianju	姚梦 YAO Meng	张明杰 ZHANG Mingjie
陈曦 CHEN Xi	黄广伟 HUANG Guangwei	林伟圳 LIN Weizhen	骆国建 LUO Guojian	王冰卿 WANG Bingqing	吴书其 WU Shuqi	许文韬 XU Wentao	尤逸尘 YOU Yichen	张进 ZHANG Jin
陈晓敏 CHEN Xiaomin	蒋西亚 JIANG Xiya	林治 LIN Zhi	宁凯 NING Kai	王健 WANG Jian	吴婷婷 WU Tingting	徐晏 XU Yan	于晓彤 YU Xiaotong	张楠 ZHANG Nan
陈修远 CHEN Xiuyuan	焦宏斌 JIAO Hongbin	刘晨 LIU Chen	彭蕊寒 PENG Ruihan	王琳 WANG Lin	夏侯蓉 XIAHOU Rong	杨天仪 YANG Tianyi	岳海旭 YUE Haixu	张强 ZHANG Qiang
程斌 CHENG Bin	李天骄 LI Tianjiao	刘芮 LIU Rui	单泓景 SHAN Hongjing	王曙光 WANG Shuguang	谢锡淡 XIE Xidan	杨悦 YANG Yue	查新彧 ZHA Xinyu	郑伟 ZHENG Wei
高翔 GAO Xiang	廉英豪 LIAN Yinghao		刘思彤 LIU Sitong					

艾心 AI Xin	方飞 FANG Fei	姜澜 JIANG Lan	刘垄鑫 LIU Longxin	沈佳磊 SHEN Jialei	王晓茜 WANG Xiaoqian	徐一品 XU Yipin	张豪杰 ZHANG Haojie	张学 ZHANG Xue
曹阳 CAO Yang	冯琪 FENG Qi	蒋建昕 JIANG Jianxin	刘晓君 LIU Xiaojun	沈珊珊 SHEN Shanshan	王峥涛 WANG Zhengtao	徐亦旸 XU Yiyang	张洪光 ZHANG Hongguang	赵伟 ZHAO Wei
陈嘉铮 CHEN Jiazheng	顾聿笙 GU Yusheng	蒋佳瑶 JIANG Jiayao	刘泽超 LIU Zechao	施成 SHI Cheng	王子珊 WANG Zishan	杨益晖 YANG Yihui	张靖 ZHANG Jing	种桂梅 CHONG Guimin
陈立华 CHEN Lihua	胡珊 HU Shan	蒋靖才 JIANG Jingcai	柳纬宇 LIU Weiyu	宋春实 SONG Chunya	吴结松 WU Jiesong	杨肇伦 YANG Zhaolun	张欣 ZHANG Xin	周明辉 ZHOU Minghui
陈祺 CHEN Qi	黄丽 HUANG Li	蒋造时 JIANG Zaoshi	吕秉田 LV Bingtian	宋富敏 SONG Fumin	吴松霖 WU Songlin	于慧瑛 YU Huiying	赵婧靓 ZHAO Jingliang	周贤春 ZHOU Xianchun
程思远 CHENG Siyuan	黄凯峰 HUANG Kaifeng	黎乐源 LI Leyuan	缪姣姣 MIAO Jiaojiao	拓展 TUO Zhan	谢星宇 XIE Xingyu	余星凯 YU Xingkai	周剑晖 ZHOU Jianhui	周洋 ZHOU Yang
迟海韵 CHI Haiyun	贾福龙 JIA Fulong	李若尧 LI Ruoyao	彭丹丹 PENG Dandan	王敏姣 WANG Minjiao	谢忠雄 XIE Zhongxiong	张本纪 ZHANG Benji	张黎萌 ZHANG Limeng	邹晓蕾 ZOU Xiaolei
崔傲寒 CUI Aohan	江振彦 JIANG Zhenyan	李文凯 LI Wenkai	邵思宇 SHAO Siyu	王却奁 WANG Quelian	席弘 XI Hong			

曹永青 CAO Yongqing	董晶晶 DONG Jingjing	黄子恩 HUANG Zien	梁庆华 LIANG Qinghua	聂柏慧 NIE Baihui	王浩哲 WANG Haozhe	文涵 WEN Han	徐新杉 XU Xinshan	臧倩 ZANG Qian
陈硕 CHEN Shuo	高祥震 GAO Xiangzhen	季惠敏 JI Huimin	刘刚 LIU Gang	戚迹 QI Ji	王丽丽 WANG Lili	吴帆 WU Fan	徐雅甜 XU Yatian	张馨元 ZHANG Xinyuan
陈思涵 CHEN Sihan	葛嘉许 GE Jiaxu	蒋玉若 JIANG Yuruo	刘江全 LIU Jiangquan	裘嘉珺 QIU Jiajun	王姝宁 WANG Shuning	吴家禾 WU Jiahe	杨瑞东 YANG Ruidong	章程 ZHANG Cheng
陈欣冉 CHEN Xinran	耿蒙蒙 GENG Mengmeng	金璐璐 JIN Lulu	刘宣 LIU Xuan	芮丽燕 RUI Liyan	王坦 WANG Tan	吴桐 WU Tong	杨喆 YANG Zhe	赵霏霏 ZHAO Feifei
陈妍 CHEN Yan	宫传佳 GONG Chuanjia	李鹏程 LI Pengcheng	刘姿佑 LIU Ziyou	苏彤 SU Tong	王婷婷 WANG Tingting	吴峥嵘 WU Zhengrong	于明霞 YU Mingxia	赵媛倩 ZHAO Yuanqian
程睿 CHENG Rui	桂喻 GUI Yu	李恬楚 LI Tianchu	娄弯弯 LOU wanwan	孙鸿鹏 SUN Hongpeng	王一侬 WANG Yinong	谢灵晋 XIE Lingjin	于昕 YU Xin	朱鼎祥 ZHU Dingxiang
从彬 CONG Bin	郭嫦嫦 GUOChangchang	李伟 LI Wei	马亚菲 MA Yafei	汤晋 TANG Jin	王永 WANG Yong	熊攀 XIONG Pan	袁一 YUAN Yi	朱凌峥 ZHU Lingzheng
代晓荣 DAI Xiaorong	黄陈瑶 HUANG Chenyao	李潇乐 LI Xiaole	梅凯强 MEI Kaiqiang	童月清 TONG Yueqing	王照宇 WANG Zhaoyu	徐沙 XU Sha	袁子燕 YUAN Ziyan	

曹舒琪 CAO Shuqi	付伟佳 FU Weijia	何志鹏 HE Zhipeng	李雨婧 LI Yujing	陆恒 LU Heng	邱嘉玥 QIU Jiayue	王秋锐 WANG Qiurui	徐雅静 XU Yajing	杨颖萍 YANG Yingping	张彤 ZHANG Tong
陈安迪 CHEN Andi	顾方荣 GU Fangrong	贺唯嘉 HE Weijia	李子璇 LI Zixuan	罗羽 LUO Yu	宋宇玛 SONG Yuxun	王熙 WANG Xi	徐瑜灵 XU Yuling	杨泽宇 YANG Zeyu	张园 ZHANG Yuan
陈辰 CHEN Chen	顾妍文 GU Yanwen	胡慧慧 HU Huihui	刘晨 LIU Chen	吕童 LV Tong	孙磊 SUN Lei	王晓坤 WANG Xiaokun	薛鑫 XUE Xin	张翱然 ZHANG Aoran	章太雷 ZHANG Tailei
陈雪涛 CHEN Xuetao	郭超 GUO Chao	黄煜 HUANG Yu	刘稷祺 LIU Jiqi	马耀 Mao Yao	孙雨泉 SUN Yuquan	王瑜 WANG Yu	杨丹 YANG Dan	张春婷 ZHANG Chunting	赵惠惠 ZHAO Huihui
陈仲卿 CHEN Zhongqing	郭金未 GUO Jinwei	黄追日 HUANG Zhuiri	刘晓倩 LIU Xiaoqian	糜泽宇 MI Zeyu	谭明 TAN Ming	王智伟 WANG Zhiwei	杨华武 YANG Huawu	张晗 ZHANG Han	赵中石 ZHAO Zhongshi
程惊宇 CHEN Jingyu	郭硕 GUO Shuo	孔颖 KONG Ying	刘信子 LIU Xinzi	宁汇霖 NING Huilin	万璐依 WAN Luyi	夏凡琦 XIA Fanqi	杨蕾 YANG Lei	张明 ZHANG Ming	周钰 ZHOU Yu
董素宏 DONG Suhong	韩旭 HAN Xu	李江涛 LI Jiangtao	刘洋宇 LIU Yangyu	潘璐梦 PAN Lumeng	王佳倩 WANG Jiaqian	谢军 XIE Jun	杨青云 YANG Qingyun	张培书 ZHANG Peishu	
方柱 FANG Zhu	何劲雁 HE Jinyan	李汶淇 LI Wenqi	刘怡然 LIU Yiran	秦勤 QIN Qin	王坤勇 WANG Kunyong	徐亭亭 XU Tingting	杨淑婷 YANG Shuting	张钤 ZHANG Qian	

图书在版编目（CIP）数据

南京大学建筑与城市规划学院建筑学教学年鉴. 2017–2018 / 王丹丹编. -- 南京：东南大学出版社，2019.3
ISBN 978-7-5641-8300-4

Ⅰ. ①南… Ⅱ. ①王… Ⅲ. ①南京大学—建筑学—教学研究—2017-2018—年鉴 Ⅳ. ①TU-42

中国版本图书馆CIP数据核字（2019）第024945号

南京大学建筑与城市规划学院建筑学教学年鉴2017—2018
NANJING DAXUE JIANZHU YU CHENGSHI GUIHUA XUEYUAN JIANZHUXUE JIAOXUE NIANJIAN 2017–2018

编 委 会：	丁沃沃　赵　辰　吉国华　周　凌　王丹丹
装帧设计：	王丹丹　丁沃沃
版面制作：	陈予婧　李舟涵　李宏健　周子琳
参与制作：	颜骁程　陶敏悦
责任编辑：	姜　来　魏晓平

出版发行：	东南大学出版社
社　　址：	南京市四牌楼2号
出 版 人：	江建中
网　　址：	http://www.seupress.com
邮　　箱：	press@seupress.com
邮　　编：	210096
经　　销：	全国各地新华书店
印　　刷：	南京新世纪联盟印务有限公司
开　　本：	787 mm×1092 mm　1/20
印　　张：	10.5
字　　数：	433千
版　　次：	2019年3月第1版
印　　次：	2019年3月第1次印刷
书　　号：	ISBN 978-7-5641-8300-4
定　　价：	75.00元

本社图书若有印装质量问题，请直接与营销部联系。电话：025-83791830